GREEN MEDICINE

GETTING WELL
WITH NATURAL TREATMENT

C.P. NEGRI, OMD, CCH, NMD

Green Medicine—Getting Well with Natural Treatment
By C.P. Negri, OMD, CCH, NMD
First Edition

Copyright © 2009 C.P. Negri

Printed in the United States of America.

ISBN: 978-0-9819884-0-5

Care has been taken to confirm the accuracy of the information presented in this work. While some data and procedures are generally accepted practices in natural medicine, many are not accepted in the practice of mainstream medicine and do not represent the consensus of medical opinion at this time. The reader should be aware of this and the author and publisher are not responsible for errors or omissions, or for the consequences from the application of the information in this book and make no warranty, express or implied, with respect to its contents. In particular, the author has made every effort to insure that the selection and dosage of natural substances in this book is in accordance with current recommendations. However, in view of ongoing research and the increasing understanding of the actions of natural substances and their interactions with other medications, the reader is urged to thoroughly research such choices before administration, especially when the recommended substance is an infrequently employed one, or totally new to the practitioner. It is the responsibility of each provider to ascertain the current status of, and possible complications from, any and all material that are dispensed or recommended.

CONTENTS

GREEN MEDICINE

GETTING WELL WITH NATURAL TREATMENT

Welcome to GREEN MEDICINE. There. Now that I've got your attention, I can tell you that the real title of this book is GETTING WELL WITH NATURAL TREATMENT. I was badgered into that new title because, as all my associates have pointed out, everyone is "going green" today.

It's great that people are finally doing something about environmental issues, eating healthier food, and looking into natural medicine—even if some of them are doing it just to be "trendy". For myself, my practice has always been "green", and 2008 marked my thirty-first year in natural medicine.

This is a primer for understanding how to get well. At our clinic, **Infinity Health Care, Ltd.**, we use everything at our disposal to help you solve your health problems. If you will spend some time with this book, you will have a better grasp of the process that our patients go through, and what you can do to get this process going for yourself. Even if you live far from a provider of natural therapies, you are not helpless. With a proper understanding of a few principles, you can start becoming healthier almost immediately.

First, please understand that even though your symptoms are probably the reason you are seeking treatment, **the symptoms themselves are almost never the problem or illness.**

Three examples of this:

1. You get a fever because the enzymes that are carried on your white blood cells can gobble up invading microbes faster when they are heated. The fever is not part of the problem; it is part of the **cure**.

2. Someone can have high blood pressure because: (a) A normal amount of blood is forced through arteries that are narrowed through the build-up of plaque on their walls (result of poor diet), or (b) Tightening of the vessel walls as a result of stress-causing excess adrenalin every day (result of poor coping habits), or (c) Mineral deficiencies causing swelling of the tissues around the blood vessels and loss of elasticity of the vessels themselves, or even (d) Inadequate intake of water has made the blood too thick. Each of these cases is different. The blood pressure is an **indicator**, not a disease.

3. Someone can have depression because certain inner needs conflict with each other and they seem impossible to resolve, causing the person to experience emotional, and even mental, exhaustion.

1

You can suppress the fever with medicines, artificially lower the blood pressure, and drug the extremes out of emotions, but making the symptoms go away has no true effect on the real illness. **It is possible for a person to be made to feel better, while actually becoming sicker.** Perhaps you have been one of these people.

Part of the problem is the conventional medical paradigm that has been long established. Discussion of health and medical matters in all media—magazines and newspapers, TV and radio, even conversations between friends—has been colonized by the orthodox medical profession. I use the term "colonized" because it is exactly like a country with a powerful army that takes over and makes another country a colony of its empire. In a similar way, other ideas of how to treat disease (and long histories of doing just that) have been replaced, just like people of colonized countries are forced to adopt the conquering country's language.

In the "language" of the common view of health and disease, a military metaphor is always used. Whenever you read or hear of a medical breakthrough or a treatment of some kind, there is usually talk of "killing the germ", or "fighting the disease". Drugs are used that are "lethal" to abnormal cells. Science is fighting a "war" on cancer. Those with the disease are described as "victims". Those who succumb to it invariably are said to have "lost a brave battle" against it. Disease organisms are "tiny foes". Even the immune system is a "guard dog" that "goes after" disease cells. Small wonder that everyone thinks that they have to "fight" a fever with acetaminophen or similar drugs.

And now we have genetic explanations of many diseases. Genetic markers are a "ticking time bomb" that can "wreak havoc", like something left in a hidden place by a terrorist. A major science magazine went so far as to use a cover illustration of the DNA double helix with a stick of dynamite strapped to it.

The result of all this imagery is that it is assumed that the processes of nature are not to be trusted, and that our bodies are nothing more than machines, like a CD player that stops suddenly in mid-song; a consumer product with a defect. When that gene kicks in, we can't hide. The immune system "turns against the body". Only the most up-to-date science (not improved living conditions or cleansing of the body) can help us "win the fight" against the disease.

Something from outside myself has done this to me, the average person thinks, *and the solution also comes from the outside.* This is the reasoning of the typical person now, thanks to years of only one picture of medicine being presented.

The popular press makes cultural heroes out of surgeons who deal with the end results of disease, fearlessly hacking away at diseased organs and tumors that might end the life of the patient. Virologists and developers of new drugs are described as "prominent researchers" and "advanced experts"; their experiments are "breakthroughs that might lead to eradication" of a disease. Genetic scientists' work is "unraveling the mysteries" and is "very promising". While all this may be true, it is implied that *only* medical authorities with the endorsement of the academic world, the government, or private industry can be trusted with an answer to disease. Only credentialed American Medical Association members, or scientists given grants in major research institutions, will be allowed to claim the "breakthroughs". No one from outside this sector is ever recognized as contributing to the "conquering" of illness, even if has been common knowledge for decades—certainly not the major innovators of natural medicine. But I ask you right now: Can you name the person who designed the first mechanical heart? It was Paul Winchell, the former TV ventriloquist I watched on Saturday mornings in my childhood (perhaps you did, too). What is not common knowledge is that many of the great breakthroughs in medicine have come from people who were not doctors—something the medical establishment does not publicize often!

Other concepts about health and disease are held by holistic approaches to medicine and by specific schools of treatment such as Naturopathy, Homeopathy, Chiropractic, and Oriental medicine. In fact, every other approach except the dominant (Allopathic) school of medicine in the U.S. has a completely different viewpoint about what is going on. Rather than a war being waged inside the body, it is simple recognition of the cause-and-effect nature of diet, lifestyle, use of the body and mind, etc. The laws governing health, and the treatments that restore it, have been well researched and established for many decades, in fact centuries, in some cases. They have not, as you are led to believe, been replaced with more accurate modern theories. They have just been put aside by regular medicine.

These concepts are largely ignored by the major newspapers, magazines, and TV shows. When they are mentioned, more often than not it is with a condescending attitude. An article may tell about the effectiveness of acupuncture for a particular ailment, but the title will be something like "Acupuncture: Trick or Treatment?". The negative headline limits anything positive the article has to say. If a natural medicine is highlighted in a story, there is a rush to emphasize that it "should not be used in place of a conventional medication", no matter if it is many times more reliable and is free of the side effects of the other drug. And why not? The major advertisers of all media are pharmaceutical companies. Count the drug ads in the next issue of a magazine you read, or the commercials during the next hour-long TV show. Don't you think it's in the best interests of a huge industry to keep people ignorant of the competition?

And there are more obviously biased media stories. "Quacks" peddle "false hope" to sufferers, unlicensed practitioners "prey" on patients, and when an alternative doctor is particularly successful, he or she is called a "so-called doctor" or "self-appointed healer". At the very best, that doctor's methods will be said to be "essentially harmless"—indicating that even if it isn't killing people, it isn't helping, either.

If you have purchased this book, you probably have already seen through this kind of propaganda. But sadly, most people haven't. Even if you know better, though, it is still important to train yourself out of the "military" ideas about medicine you probably grew up with. Stop using words like "fight", "kill" "battle", "conquer", and like terms to describe the reversing of illness. Language has a bigger impact on how we think than we realize.

Much of this imagery has come from the germ theory. Ever wonder why they call it the germ "theory"? That's right. Because, as impossible as it seems, it has never been proven to be completely true. In order to be true, it must fulfill all of what are called "Koch's Postulates". Basically, a sick person should be found to have an organism growing inside him that can be isolated. The organism is taken from the body, allowed to grow under correct conditions, and then put into a healthy host's body. Then the host should manifest the same disease as the first person from whom the organism came. It might strike you as ridiculous to even question this straightforward approach.

The problem is that most "bugs" are not so accommodating. People harboring the same germ don't always develop the same illness. Sometimes they don't get sick at all. Diseases that are supposedly caused by a particular organism will also occur in someone who was not exposed. As an extreme example, there are people with AIDS who tested negative for the HIV virus. There are also people who are HIV positive who never develop AIDS. Unless you are an inquiring type, you will likely never come across these facts. Needless to say, this embarrassing reality for the medical community is downplayed in its conversations with the public.

The germ theory became popular because at the point in time it took hold, it fit in with the other generally mechanistic theories in science. But it continues to be popular because it enables us to avoid responsibility—the cause for disease is outside one's self. "*I* didn't do anything wrong; this *germ* picked on me." And it is a great boost to commercial interests because in order to fend off this vast "army" of disease-causing microbes, we need vaccines, drugs, disinfectants, and any "new, improved formula" products that continue to be developed and sold to us.

Scientists realize that this view of the interaction between the sick person and the germ is terribly oversimplified and simply does not fit the facts of disease very often. Nevertheless, a huge industry depends on most people embracing this idea of how things work. There is a large body of research work dating back decades that shows bacteriologists proved that microbes change from benign to pathogenic—friendly to harmful—when the internal terrain of the body is changed. But this discovery did not change the standard textbooks*.

Without going into all the complicated scientific details of these discoveries in bacteriology, the matter can be summed up quite simply: **Microbes are a factor in causing disease, but disease is a factor in creating microbes**. Or to put it another way, germs don't necessarily cause disease, disease causes germs! If one recognizes this seemingly backwards idea, it suddenly becomes clear why people were being cured of every kind of infectious disease at Naturopathic clinics where diet, applications of water, detoxifying, and herbs were being used—before antibiotics were widely available. Even after the new "wonder drugs" were common, these "nature cure" sanatoriums were famous for restoring health when they didn't offer one germ-killing therapy. By changing the internal terrain of the body with the clearing of toxic waste and building up of the tissue with concentrated nutrition, the microbes living there no longer had a hospitable environment. Even conventional medical facilities used such methods, such as the old tuberculosis hospitals before the

4

antibiotic age. Fresh air, lots of sunlight, and better nutrition cured many thousands of people before a single dose of Streptomycin was ever given. So this concept does not belong exclusively to alternative medicine; conventional medicine has just mislaid it, so to speak.

* Some of the published works detailing the variability of microbes include:

Kendall, A.I.; Rife, R., "Observations on Bacillus Typhosus in its Filterable State", *California and Western Medicine,* Dec. 1931

Rosenow, E.C., "Transmutations Within the Streptococcus-Pneumococcus Group", *Journal of Infectious Diseases,* Vol. 14, 1914

Rosenow, E.C., "Observations on Filter-Passing Forms of Microorganisms", *Proceedings of the Staff Meetings of the Mayo Clinic,* July 13, 1932

Rosenow, E.C., "Observations With the Rife Microscope on the Filter-Passing Forms of Streptococcus", *Science,* Aug. 26, 1932

Hume, D.E., *Bechamp or Pasteur?,* C.W. Daniel Co., Essex, 1947

Sonea, S.; Panisset, M., *A New Bacteriology,* Jones & Bartlett, Boston, 1983

Domingue, G.J., *Cell-Wall Deficient Bacteria,* Addison-Wesley, Reading MA, 1982

To compare with the germ theory, I have contrasted it with what we might call the "host theory", where the internal conditions of the body determine the nature and actions of the germs.

TWO THEORIES OF INFECTIOUS DISEASE

GERM THEORY	HOST THEORY
Disease is caused by microorganisms outside the body	Disease results from microorganisms within the body changing
To prevent disease we have to defend against the microorganisms	To prevent disease we have to generate health
Microorganisms have one form (monomorphism)	Microorganisms change color and shape depending on what they are feeding on (pleomorphism)
A disease is associated with a specific microorganism	A disease is associated with a specific condition inside the body
The function of a microorganism is constant	The function of a microorganism changes when the host tissues are injured from chemical, mechanical, or nutritional insults
The microorganism is the primary cause of disease	The microorganism becomes disease-causing as the condition of the host breaks down

In other words, disease occurs when conditions (inside or outside the body) are favorable for the microorganisms to acquire the qualities that allow them to do negative things. Remember, what many medical doctors today call "biological medicine" or "functional medicine" is based on this concept, and it is that same concept that has been central to natural or Naturopathic medicine for many, many years. It is not a view held by a small minority, but in fact has been upheld by some of the greatest names in science.

I wrote a short article some years ago that I think simplifies the confusing matter of health and disease. I am including it here.

COMMON SENSE HEALTH...AND AUTO MECHANICS

I had a problem with my beloved little car recently. When I started it up in the morning, it was running pretty rough. The night before, it was fine. Having never had any trouble with my car before (at 135,000 miles!), I felt a little betrayed. What could be causing it?

A mechanic said many things could cause it. While he was checking them out, I did a mental review of systems and it wasn't long before I figured out how to help the average person make sense of the human body.

When you get sick, it's easy to feel betrayed by your body. After all, isn't it supposed to maintain itself? Well, yes—to a point. But you wouldn't expect your car to run forever without maintenance, would you?

A car that was jerking all over the road like mine did could have several things wrong. Same for the human body.

<u>Could it be something that got in the fuel tank</u>? Cheap quality gas, condensation in the tank, or particles from the breakdown of the lining of the tank could cause problems with the engine.

The "fuel tank" in the body is the digestive system. Give it bad fuel, or irritate it over time, and changes in the lining will cause your "engine" to run poorly. These changes can even cause you to not absorb the nutrients even when you are eating good-quality "fuel".

<u>Could it be something wrong with a filter</u>? A fouled-up fuel filter will stop a car flat. Fuel filters in a human include the liver, kidneys, and lymphatic system.

If the liver is over-burdened with too many impurities absorbed through eating, breathing, and touching toxic substances, it will not filter the blood efficiently. And if the blood is not carrying enough nutrients in the first place, a burden is placed on your engine.

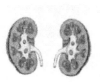

The kidneys have to filter the blood also, and these filters are burdened by too many impurities generated by the body itself. When you burn fuel, there is always a byproduct. Too much byproduct clogs a filter. You probably clear hair out of your bathroom sink drain regularly, but did you <u>ever</u>, even <u>once</u>, think about doing something for your kidneys?

7

The lymphatic system is a huge network of vessels that carry lymph throughout the body, trapping and neutralizing viruses and bacteria to keep you from getting sick. Ever wonder why you don't catch <u>every</u> one of the hundreds of infections you're exposed to every day? Thank your lymph vessels and your spleen, for making antibodies that fight invading illnesses. That is, if the spleen and lymph are doing their jobs.

But wait—maybe your engine isn't running smoothly because of dirt in the <u>air filter</u>.

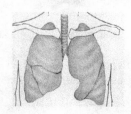

Those wonderful air filters called the lungs don't just catch impurities from the outside; they are part of your body's "carburetor". They insure that the blood is rich in probably the most important nutrient of all—oxygen. If oxygen saturation of the body is low, the amount of carbon dioxide increases. If your car's exhaust was bad, and the inside filled up with carbon monoxide gas, you could soon be dead! But if you drove short distances with the window cracked, you would just get sicker over time without ever knowing why.

Part of the body's "exhaust" is carbon dioxide gas. If you don't get enough oxygen, and you don't get rid of enough CO_2, you slowly get sick and don't know why.

And speaking of the exhaust system, your car won't run if you plug up the exhaust pipe. Your colon is the biggest exhaust pipe in your body. How can you expect your body to run if the colon is partially plugged up?

Even if you have regular bowel movements, deposits could be sticking to the walls of the colon, preventing absorption of minerals and actually recycling poisonous waste. Creating health always depends on a well-functioning colon.

<u>Could it be the electrical system?</u> In modern cars, the electrical system is very important. In my car's case, as it turns out, there was a worn-out distributor wire and one bad spark plug. In the body, the nerves and the peri-neural (meridian) system is the electrical system.

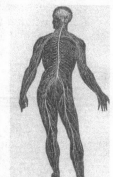

The nerves are the more dense electrical "wires" and the meridians are the finer ones, and without these tiny amounts of electricity going everywhere, the organs do not receive all the signals telling them what to do. Even the cells themselves rely on electrical impulses to behave normally. Abnormal cells start to form when chemical changes around the cells cause the electrical impulses to change.

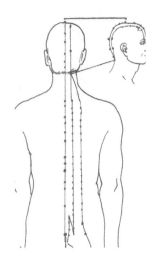

This is why it is so important to keep the electrical system running smoothly. In the Far East, people have relied for centuries on regular acupuncture treatments to keep their electrical systems running smoothly. Here in the West, we give our cars regular tune-ups but <u>not</u> <u>our</u> <u>bodies</u>. Why?

The first thing I did when I picked up my car at the garage was point it straight and drive it without touching the steering wheel. I wanted to see if it ran straight. If it didn't, it could be that a tire was under-inflated or worn-out, or my alignment could be off.

The joints in the human body need alignment, too. If the muscles, tendons, and ligaments do not have the right tone, movement causes your joints to work in an uneven fashion. Over time, the joints can be damaged, and at the very least, inefficient movement places a bigger burden on your internal organs.

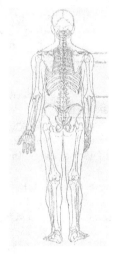

This goes beyond your body itself. What about your clothes and furniture? Everyone knows how high-heeled shoes put a strain on the lower back. But how about clothing and furniture that causes your body to perform inefficiently? How about a chair that causes you to slouch or a young person's backpack, both of which interferes with good spinal alignment and make it harder to breathe deeply and harder for the heart to beat? Children used to be scolded for poor posture but today that's seen to be an idea from a more formal and severe time. The need for good posture and efficient movement is a matter of <u>health</u>, not style! "Charm school" graduates years ago were actually learning how to use their bodies in a healthy manner. Getting well requires more attention to effective lining up of the body.

Now these are all pretty much common sense—the workings of both automobile and human body. If there are so many things that can cause the engine to not run smoothly, and if the "engine" of the human body is the heart, <u>is it any wonder why</u> <u>heart disease is such a common killer</u>? It's because anything can ultimately cause it, from bad shoes to bad food, from poor breathing to poor thinking.

Now if you took your complaint of an uneven-running engine to a conventional doctor, you would likely get a report like this:

We cut a sample of metal from your car and examined it. We believe there were impurities in the metal during the smelting process, causing the part to later buckle, causing malfunction. This may be the case throughout the car. We want to cut out a few more pieces of metal and examine them. If we find flaws, we will replace all the metal. Metal made today is better than when your car was made. Anyway, the problem can only be seen under laboratory conditions, so don't bother looking yourself.

This is not to say that conventional physicians are not any good at what they do. But they work in a system that primarily trains <u>disease specialists</u> rather than whole-body health promotion, and tells them to use invasive, segmented, and toxic methods first. Most will consent to gentle, natural methods as a <u>last</u> resort. To me, this is backwards. The least dangerous methods should be given a try first, then stronger methods if those fail. Procedures that work well for emergency medicine are poor for chronic disease.

Don't think that I'm advocating just eating lots of fruit, thinking nice thoughts and waiting a long time to see results. I believe that normal body systems must be stabilized before looking for more obscure causes of an illness. Most of the time, if nutrition, filtration, elimination, electrical function and alignment are brought to within normal range, the problem clears up.

It doesn't make much sense to me to treat some problem directly while leaving several major body functions barely operating. If you are treated that way, you might see some changes in symptoms, but you won't be healthier.

I assume that people come to me to become healthier, not just to have symptoms treated. To work together to do that, I ask for their cooperation. In return, I promise to do things that **make sense**.

FIVE FUNDAMENTAL CAUSES OF ILLNESS

1. **Excesses** in the body: Toxic substances, or too great a quantity of non-toxic substances (excessive intake of food, etc.).
2. **Deficiencies** in the body: Absence of elements needed for functions, growth, maintenance, or repair of tissues.
3. **Stagnation**: An interruption of the normal distribution of elements that modify or maintain functions, or repair tissues.
4. **Environmental causes**: Having to live or work in surroundings that are either directly toxic or place a greater stress on the person than can be adapted to.
5. **Mental / emotional causes**: Problems that stem from difficulties in coping with life situations, and which impact on the body's functions. These can also be excesses, deficiencies, or stagnations in nature. For example: excess fear is unhealthy, but a complete absence of fear (deficiency) will lead to trouble also.

Physical trauma or injury causes stagnation, which then causes an excess, a deficiency, or both. Emotional trauma is no different. There are few physical traumas that do not cause emotional traumas, and few emotional problems that will not cause physical symptoms in time.

So the questions we must ask are: Where is stagnation? What is excessive? What is deficient? How can I better deal with my environment? How can I better deal with myself?

Since beginning my training in 1974, I have observed many different methods and medical philosophies. Sorting out failures and triumphs, I evolved my "rules of effective care", which are followed by my associates and myself. We started calling them "Negri's Rules", but to my mind they ought to be just common sense.

NEGRI'S RULES

1. There is almost <u>never</u> only <u>one</u> cause of a problem or disease. Look at all factors.
2. Do not depend on only one form of diagnostic testing to define the problem.
3. Do not depend on only one form of treatment if you want reliable results.
4. Treat all levels: physical structure, chemical, electrical, subtle, mental and emotional.
5. Use many complementary methods that enhance each other without conflicting.
6. Stabilize the condition while working on the root causes.
7. Treat the common factors first, moving toward the individual factors.
8. Treat the gross processes first, moving toward the subtle.
9. Gentle, non-invasive, non-toxic therapies should be the <u>first</u>, not last, resort. Use stronger methods as required.

WHAT KINDS OF THINGS DO WE TREAT?

PMS

Asthma

Menopause Symptoms

Ear Infections

Prostate

Depression

Family Health Maintenance

Allergies

Anxiety

Reflux

Arthritis

Headaches

Neck and Back Pain

Fibromyalgia

Irritable Bowel Syndrome

Yeast Infections

...IN SHORT, JUST ABOUT EVERYTHING!

YOUR HEALTH QUOTIENT

You are the sum total of everything that you have eaten, everything you have done, everything you have thought. All your experiences and all the substances you come in contact with on any given day are recorded within your body-mind system.

More than that, you are the end result of generations of genetic information. You are not only who *you* are, but you are—at least partially—who your parents were and who theirs were. That picture of Grandma when she was young that looks so much like you? That's not just coincidence. Besides a certain look about the eyes, there may be predispositions to certain illnesses that you share with your grandmother. But your health history may differ from hers because of different things that happen during your lifetime.

You come into this world with your own biological makeup, but within that is genetic information from your ancestors. You are not very old before a doctor injects some vaccines into you and implants new information on top of that. New proteins—not even from a human, but from monkeys and pigs—are adopted by your body chemistry. Some of these cause only the mildest response internally; others may turn your health in a new direction.

You are also developing mentally at this point in time. Suppose you are crying and your mother does not come right away. You cry harder. Why doesn't she come? If this happens enough times, the distressing memory is firmly deposited in your psyche and you may develop a personality that is clinging and irrationally afraid of being abandoned. Everyone will think it's just "who you are", even if your brothers or sisters have very different personality traits.

Going through the normal childhood diseases trains your developing immune system to recognize and fight disease[*]. One of these self-limiting illnesses may have left a toxin that will continue to have an impact on your health even after you grow up. A toxin is a poisonous residue that the body recognizes as being something to eliminate. Sometimes, it does not succeed and the toxin will deposit and become part of the general information of the body.

As you mature, you may develop food sensitivities. Some of these will be too subtle to cause a rash or anything that you would notice, but these problem foods will not digest completely and they will leave remains that are toxic to the system. In trying to deal with that, your body expends quite a lot of energy, sometimes creating an inflammation that does not seem to have anything to do with food or digestion.

As you go through life, you are exposed to increasing amounts of chemicals, electromagnetic fields, metallic toxins, and of course more viral and bacterial illnesses. Some of these will be dealt with efficiently by your immune system; others

---

[*] This is one reason many medical authorities feel that vaccination for self-limiting childhood illnesses is not a good idea. The developing immune system is circumvented, by stimulating it with a fake illness rather than the real one that actually strengthens the child.

may make an impact that keeps them on the "roster" of your total body information. The more toxic residue that deposits in your tissues, the more energy your body uses every day to try to shake them loose. That leaves less energy to fight off new toxins. An illustration of this is on the next page. This chart will show how your body and mind are simply a collection of data, put there both before birth and as you move through life.

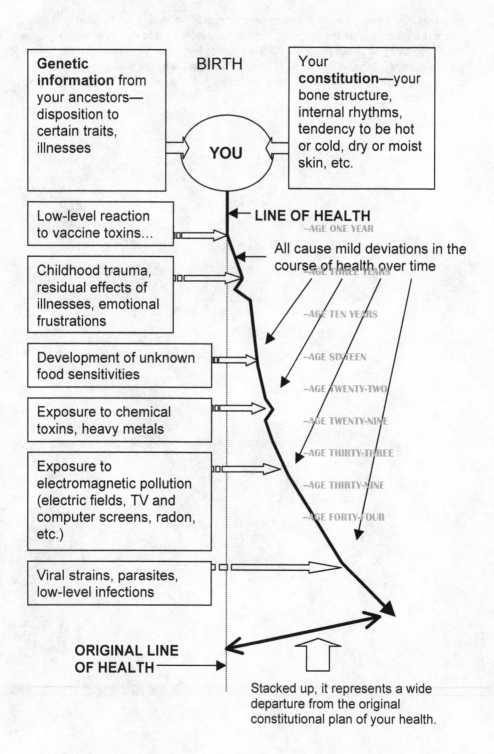

Genetic information from your ancestors—disposition to certain traits, illnesses

BIRTH

Your **constitution**—your bone structure, internal rhythms, tendency to be hot or cold, dry or moist skin, etc.

YOU

Low-level reaction to vaccine toxins...

LINE OF HEALTH

--AGE ONE YEAR

All cause mild deviations in the course of health over time

Childhood trauma, residual effects of illnesses, emotional frustrations

--AGE THREE YEARS

--AGE TEN YEARS

Development of unknown food sensitivities

--AGE SIXTEEN

--AGE TWENTY-TWO

Exposure to chemical toxins, heavy metals

--AGE TWENTY-NINE

Exposure to electromagnetic pollution (electric fields, TV and computer screens, radon, etc.)

--AGE THIRTY-THREE

--AGE THIRTY-NINE

--AGE FORTY-FOUR

Viral strains, parasites, low-level infections

ORIGINAL LINE OF HEALTH

Stacked up, it represents a wide departure from the original constitutional plan of your health.

16

DETOXIFICATION AND DRAINAGE

Scientific natural medicine emphasizes the role of toxins in disease, and centers on detoxification as the main method to get well. But toxins can exist in many forms, and the accumulation of toxic elements and trauma are what lead to most chronic illnesses. Trauma and toxic substances tend to pile up, causing layers to form as you get used to each new level of sub-health (we call these "tolerance levels"). When some of these elements have combined to make you sick, your body-mind system must be stimulated to identify them, neutralize them, release them (detoxification) and expel them (drainage).

It is not an "alternative" medical idea that health problems result from the accumulation of foreign substances in our bodies. Exposure to environmental toxins has reached disturbing levels. Millions of chemicals are used commercially every day and we are exposed to far

---

What Are the Signs of Toxicity?

Fatigue, "brain fog", forgetfulness, headaches, bloating, gas and indigestion, alternating constipation and diarrhea, allergies, overeating, mood swings, bad breath, stiff joints, lower back pain, poor circulation, cold hands and feet, recurring infections, skin eruptions, and even cellulite are just some of the usual symptoms.

---

more than we think, through household products, our food and drink, and the air we breathe. Drugs we have taken leave residue in our bodies, also. The Environmental Protection Agency has mandated that warnings to consumers be put on more and more products. Fresh and frozen fish, for example, exposes people to mercury that can cause a large number of serious illnesses. Dentists are now being required by law to inform patients of the potential risks of silver amalgam fillings (also containing mercury).

What is the solution when we cannot eliminate all exposure to these agents? **We must increase the body's natural instinct to detoxify.** The burden is too great to expect that our bodies can eliminate poisons at the same rate as we encounter them today. A good cleansing program and clearing of toxins is essential for restoring health.

But disease is not made completely of toxic elements. All these (often microscopically small) factors, from metallic toxins and localized infections to emotional trauma and unresolved mental conflicts, will cause disturbances in the body's chemistry. These changes will lead to changes in organic functioning. In other words, illness moves from the subtle to the gross; from tiny electrical currents to the growth of tumors.

In medicine, the realm of your basic biochemistry and how it affects organ function is called your **biological terrain**. Assessing the bio-terrain is necessary to know which parts of your chemistry need to be balanced so you can heal. Just correcting the subtle factors does not always correct the gross factors. Here is the sequence in which one level of factors creates a new one:

17

```
┌─────────────────────────────────────────────────────┐
│ Chemical toxins, Heavy Metal toxins, EMF stresses,   │
│ Food sensitivities, Dietary stresses, Vaccine/drug   │
│ residuals, Mental/Emotional stresses, Inherited      │
│ predispositions, etc.                                │
└─────────────────────────────────────────────────────┘
                          │
───────────  BIOLOGICAL   ▼        TERRAIN  ───────────
┌─────────────────────────────────────────────────────┐
│ Acid-Alkaline Imbalances → Oxidation/Reduction       │
│    Patterns → Sympathetic/Parasympathetic Nervous    │
│    System Imbalances → Malabsorption → Vitamin/       │
│    Mineral Deficiencies                              │
└─────────────────────────────────────────────────────┘
                          │
                          ▼
┌─────────────────────────────────────────────────────┐
│ Low HCl and Enzyme Production; Infections, Parasites;│
│ Adrenal, Thyroid, Liver, Pancreas, Kidney imbalances;│
│ Organ and Meridian dysfunction                      │
└─────────────────────────────────────────────────────┘
                          │
                          ▼
         ┌───────────────────────────────────┐
         │ Result:  Your Symptoms            │
         └───────────────────────────────────┘
```

HERING'S LAW

Now you know that many factors combine to cause changes in body chemistry that will keep you sick unless corrected. Let's look at what happens once we turn that process around.

In all schools of medicine, there is an observation that healing takes place in three predictable directions:
1. From deep inside the body outward toward the surface;
2. From the top of the body downwards;
3. In reverse order of appearance.

This is called *Hering's Law*. Since the body's normal programming tells it to discharge toxins, when you get sick there are usually some kind of elimination symptoms (runny nose, cough, sweating, vomiting or diarrhea, etc.). That is, the body tries to push something out. If you had a chest cold, and the cough gets better and you are left with a drippy nose, you know you are getting over it—the illness is moving from the deep (the vital organs called the lungs) to the superficial (mucous membranes in the sinuses). On the other hand, if you have a runny nose and then you develop cough and lung congestion, you know instinctively that you're getting worse, right? Now you know why—illness moves from superficial to deep and from secondary organs to primary organs. Healing goes the opposite way.

Usually if you have a rash that is climbing up the body or spreading outward, it is getting worse. When it gets better it recedes. As another example: If you have pain in many joints, you will notice that as you get healthier your shoulders will improve before your elbows, elbows before knees, etc.   This is the top-downwards phenomenon. No one knows why it works that way, but it does.

The third aspect of *Hering's Law* is the rule that healing occurs in reverse order of appearance. This means that your most recent problems will be addressed by your immune system first; then what you had before that, and before that, etc. For this reason, old problems you may have had years ago can return briefly as you are healing. Some call this "re-tracing". The same thing happens in psychotherapy as well. Once the body and mind have a chance to go back over old unfinished business, they tend to want to get rid of the problem for good.

We at Infinity Health Care believe that getting better requires balancing all levels where there may be a problem. Some people whose illnesses mostly stem from electrical imbalances will get miraculously better from acupuncture. Other people with the same illness will not get completely well. They may have subtle emotional factors that keep the condition in place. These people may see dramatic results from taking gentle Homeopathic medicines. A few might not. These individuals may have chemical imbalances that require medicines that have a chemical action. Herbal medicines may work better here.   One or two of those may have nutritional deficiencies due to malabsorption; all the herbs, acupuncture, massage, or diet in the world will not fix their problem. Metabolic repair is needed.

19

We blend the many methods of natural medicine to fit your individual set of co-factors. That way, you are less likely to "fall through the cracks" than someone being treated only one way.

What's Our Approach to Treatment?

Micro-energetic Treatment to correct subtle electrical impulses that communicate with every part of the body. These influence the nerves' activity, they in turn influence blood flow. Balancing the electrical system of the body is essential for health.

Metabolic Treatment to balance the body's biochemistry. The basic foundation of
- **acid/alkaline balance**
- **absorption factors**
- **electrolyte balance**
- **enzyme pathways**
- **vitamin/mineral uptake**

are all crucial to healing and maintaining proper organ function.

Natural Medicines, little-known in this country, have been used for some time in European hospitals with great success. Homeotherapy, isotherapy, oligotherapy, gemmotherapy, as well as botanical medicines (phytotherapy) are used to restore health with none of the usual side effects of synthetic drugs. But make no mistake—just because they are gentle does not mean they are not effective!

Detoxification will target specific toxins within the body and ionically displace them from their cellular binding sites.

Drainage enhances the body's excretion systems and optimizes the elimination of toxins from the body.

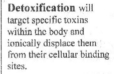

Diet is used to both enhance treatment and then maintain improvement afterward. The foods that help or harm an individual can be found by testing.

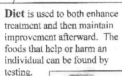

Infinity
HEALTH CARE, LTD.

THE HEALING CRISIS

There are very few treatments we use at Infinity Health Care that can cause what could truly be called "side effects". But all natural treatments can bring about a "healing crisis". This is a state where the symptoms of your problem increase temporarily as your body mounts its healing reaction against the problem. Once this is in motion, the symptoms subside as the illness becomes weaker and your vitality becomes stronger. **Most chronic health problems will have short periods where they worsen, followed by an improvement. This is natural and should not be interfered with.**

A healing crisis is the result of two mechanisms. One is the aspect of *Hering's Law* that the healing force moves backward through all the layers of illness and trauma. When it pauses to address a particular problem, those symptoms become more noticeable. Sometimes it is an old problem that stopped producing symptoms long ago. They return as the spotlight falls on the problem and it begins true healing. This is not just true of the body. When someone is undergoing counseling and overcomes an emotional problem, it is common for old feelings or memories to return. This well-known action is a feature of *Hering's Law*.

The second mechanism of a healing crisis is the release of toxins in the various tissues of the body. It is not enough to release them; they must be expelled. To do this, the liver, kidneys, lymphatic vessels, skin, bowels, and lungs all have to process the changes. Depending on your level of toxicity, there can be a "traffic jam" as the eliminative organs work overtime to expel all the released poisons from your body. This is known medically as a *Herxheimer Reaction*. The symptoms that occur when this is happening can range from a vague sluggishness to symptoms that resemble a cold or the flu, but they can be distinguished by the fact that they do not last as long as a real illness. Dizziness, aching in the joints, headache, bad breath, diarrhea, mild fever, runny nose, or skin eruptions may occur. Take note: **Don't use medicine for a rash or a discharge that appears in this way. You don't want to push a poison back inside that your body is pushing out.** Most symptoms such as these last 72 hours at the most. We always take steps with your treatment plan to insure that your eliminative ability is enhanced, so as to reduce or eliminate as much as possible any "traffic jam" discomfort.

In the case of infectious organisms, which play a role in many different types of health problems, you might think that "killing the bugs" gets rid of the problem. This misconception has caused millions of people to undergo repeated courses of antibiotics and yet still be sick. In natural medicine, we sometimes need to use antibiotic therapy and have many potent ways to fight both chronic and acute infections. Yet, we recognize that without helping your body clear the debris of the killed organisms, you will have repeated infections. I like to think of it as a neighborhood overrun with vultures. When the vultures are so thick that normal activity is interfered with, the residents get the army to come in and shoot all the birds. Once all the vultures are dead, people come out of their houses again and try to resume their lives. Is it the end of the vulture problem? With hundreds of dead vultures lying everywhere? No. Before long there will be a new crop of big, black birds circling overhead, because there's food on the ground.

Bacteria are nature's garbage collectors. They might be few in number, but when there is a good buildup of garbage in a specific place, they reproduce like crazy because there is lots of food!

This is why one day you have a normal amount of streptococcus bacteria in your mouth and throat, and the next day you can have a sore throat that develops into *strep throat*. Your throat culture will show that you are teeming with these "bugs", yet you didn't kiss anyone with a strep infection. Where did they come from? They seem to appear by magic.

We have many treatments at Infinity Heath Care that can cause healthful organisms to fight off infectious ones, and some that can actually make the bad bugs die off in short order. But we always want to clear the channels for your body to eliminate the waste material. I should add that many chronic problems have viral, bacterial, or fungal components that are missed with orthodox medicine.

If you have recurring bladder infections or yeast infections, or pelvic inflammatory disease, you may have a cleansing reaction of increased vaginal discharge. If you have chronic bowel problems, it is not uncommon to pass a worm into the toilet at some point. As repugnant as this may be, keep in mind that it is better to have a dead parasite come out than to keep a living one inside.

CYCLES IN HEALING

Healing takes place in cycles of activity just as the accumulation of illness does. You didn't get sick at the same steady pace all along and you won't get well all at once, either. **It usually takes about one month of healing for every year you've had a problem, in order for it to be truly reversed.** Along the way there will be periods where you will rapidly improve, and other times where things seem to stand still. This is natural. Everything does not have the same rate of change. By the time you have all your teeth as a child, you do not have all your bones yet. By the time you have your full adult size, you still do not have your full maturity or intelligence. Everything happens in a sequence.

Here's what happens as your health declines:

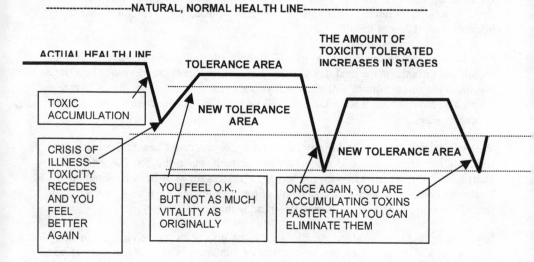

Getting healthy is just the opposite. Notice that it also happens in stages, with cycles of improvement:

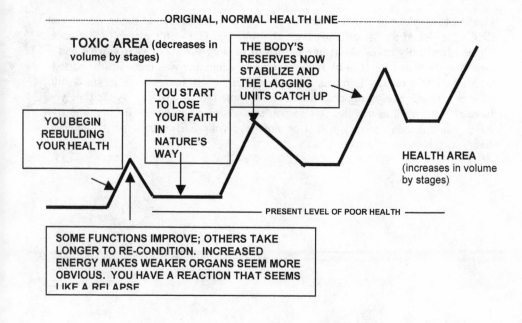

THE RULE OF SEVEN
Thankfully, it takes only a small fraction of the time to regain your health that it did to destroy it, thanks to the Rule of Seven. Nature operates under the Rule of Seven. She uses units of seven in many ways.
- The moon completes the cycle in 28 days from full moon to full moon.
- A woman's menstrual cycle is 28 days.
- A baby's gestation period is 280 days.
- Chicken eggs hatch in 21 days.
- Most ducks hatch in 35 days.

So, also it is with the human body:
- It takes 7 days to cleanse the bloodstream.
- It takes 7 weeks to cleanse the lymph glands.
- It takes 7 months to cleanse the endocrine system.
- It takes 7 years to change all the molecules of the body.

"You can't clean the house without stirring up a lot of dirt first. The house invariably looks worse before you can make it look better."

---V.E. Irons

The "Rule of Seven" was also often quoted by Henry Lindlahr, M.D., a famous doctor who raised natural medicine to a higher scientific standing, and founded an acclaimed Naturopathic medical school. But even without any medical authorities to point this truth out to us, we as a society have long recognized the fact that there are cycles of seven, especially seven-year periods, in our lives. Our most formative years are between the ages of one and seven, and society considers you fully matured after three cycles of seven years (21). Once you are legally an adult, further cycles follow. Like the important growth period of the first seven years, the years between twenty-eight and thirty-five are when we typically establish who we are going to be for the rest of our lives. Through this up-and-down seven-year period, we work out (or try to) those matters that allow us to settle into what is called "middle age".

Our illnesses also have a youth, middle age, and old age (if we live long enough). Healing likewise has distinct periods where one thing is called for, then something different, to make the process of reversing disease stay active. During that process, your natural doctor will help you navigate the twists and turns in order to stay on track. As mentioned earlier, symptoms will often worsen, then clear, then later something else will arise. I have seen this routinely, throughout decades of practice. New patients will usually report feeling better or worse in a week (7 days). They will improve over a period of almost two months, then return at eight weeks saying, "I started feeling bad a week ago…" (7 weeks). After they have improved, become much healthier, and are not bothered by the original problems they came in for, I know I will hear from them about the seven-month mark. Typically there is a "new" problem, and most of the time it is actually a deeper, unresolved condition that was not bothering them at the outset.

If you don't understand that there are natural principles at work here, then it will seem like everything that happens to you, pleasant and unpleasant, is random. It is not. When you enter puberty, it is an unsettling time. Then you go through the teenage years and get comfortable with it, only to have a new set of uncertainties as a young adult. Entering a new phase of healing is much the same. If you misinterpret the new discomfort as going backward (instead of going *forward*), you will be powerless to make that transition smooth. Many people are enthusiastic about their treatment at first, experiencing great improvements, only to discontinue later because it is not as dramatic, or something "new" pops up. Please discuss with your doctor what is happening so measures can be taken to make the healing process less "bumpy". If no one knows what you are going through, no one can help you, and you will never really know what is happening to yourself. As you saw from the previous chart, it takes a certain length of time for lymphatic problems to clear, but longer for hormonal problems. Most people have a mixed batch of conditions, so you cannot judge one problem by how quickly another cleared up.

It is important that you have an understanding of what is happening in the body during the process of restoring health. Let's look at the organs involved in your "house cleaning".

DETOXIFICATION AND THE LIVER

As the main "filter" for detoxifying the body, the liver takes toxins from your environment, food, and microbes and binds them with chemicals made inside itself. This process is called *conjugation*. Then the combined mixture is excreted through the intestinal tract. If the liver is overburdened, you don't throw off poisons well. The liver is poorly functioning long before it reaches the point of having abnormal enzyme values on a standard blood test. Did you ever notice that the most common side effect from taking synthetic drugs is liver toxicity? Taking a drug for a completely different problem results in decreased liver function. There are a number of natural medicines that help restore the liver and we use them all at Infinity.

DETOXIFICATION AND THE KIDNEYS

The kidneys also filter the blood and excrete toxins through the urine. Like the liver, the kidneys can be very poorly functioning before a problem shows on tests. When the kidneys don't do a thorough job, the skin often takes up the slack and becomes a more active organ of elimination. Toxins being pushed out through the skin are the reason for many skin eruptions and can be due to sluggish kidneys.

DETOXIFICATION AND THE LYMPH

The lymphatic vessels are an important part of your immune system. Here, cells trap and neutralize infectious and poisonous materials and neutralize them. That is why your lymph nodes swell up when you get an infection. Then the materials are slowly pushed into a place called the thoracic duct, and are burned up during the process of metabolizing food. When there is more debris in the lymph channels than it can handle, your immune function is lowered and you are carrying a larger-than-normal toxic load. There is almost five times the amount of lymph in your body as there is blood. That is a lot of fluid to keep clear and circulating. In addition to treatment, there are some simple things you can do at home to help your lymphatic system.

DETOXIFICATION AND THE SPLEEN

The spleen organ forms blood cells, filters and stores blood, and is a vital part of your immune system. Chronic and acute infections are possible if the spleen is not functioning well.

DETOXIFICATION AND THE GASTROINTESTINAL TRACT

With regard to eliminating poisonous substances once they are shaken loose from organs, connective tissue, and other sources, we have to talk about the colon because it is a large exit route out of the body. Whatever we do to clear debris from the intestines benefits us all over. Remember that most chronic health problems have some infectious organisms as co-factors.

The "holding tank" for these infectious organisms is the intestinal canal. It typically contains eight to twelve **pounds** of living organisms! Abnormal bacteria, fungi, viruses, mycoplasmas and parasites will live there and can spread to other tissues. It's amazing; the gut can have as many as 100,000 **billion** microorganisms at any one time (in 400 groups or types).

Now consider this. Since the average person has about 10,000 billion cells in his or her body, that means that there are ten times more of "them" than you!

The mucous membrane of the intestines is made up of epithelial cells, only one cell in thickness (monolayer). That's right; only one cell layer exists between your internal organs and the things that come into your body from the outside. That layer is about the size of *two tennis courts*, if stretched out flat. This tissue can become a huge warehouse for infectious organisms and toxic substances, and if the condition of this thin membrane is not good, the body can be infiltrated from the contents of the intestines. This leads to a wide variety of diseases that don't seem like they have anything to do with the GI tract.

When there is trouble in the GI tract, two different things can happen. Waste material can be poisoning you while giving a home to infectious organisms, and the poor condition of the lining keeps you from absorbing the nutrients in your food. This creates both an *excess* (toxic material) and a *deficiency* (malabsorption) at the same time.

Like other natural medical approaches, our protocols at Infinity Health Care give great attention to any possible problem in the GI tract, because this is so often the major factor in a sick person's recovery. We use a number of methods to insure good internal cleansing and to restore the ability of the intestines to function properly.

If you have only 5% kidney under-function, but 15% lymphatic congestion, 10% liver toxicity, and "only" a 20% reduction in bowel efficiency, it stacks up to <u>50% toxicity</u>!

Can you see now how almost any illness can result if several systems are not working at 100% efficiency?

I was lecturing at a medical school one time when a senior doctor there said, "Toxins don't build up in the body!" I said, "Then why do you put patients with kidney failure on dialysis?" He turned and left the room.

Another time I was confronted with the statement that the colon "cleanses itself naturally". Not only the huge number of sales of laxatives in this country should contradict that, but the rising number of colostomies performed for blocked bowel should end all argument on this point.

We must face the fact that medical politics determines what we learn and how we think about health. Simple methods such as detoxification will never generate as much revenue as controlling disease through drugs. Therefore, many arguments are made against natural procedures in a desperate attempt to save face or to discredit the unorthodox source (such as myself). Some of these arguments are laughable; others take some thought to realize just what they are—propaganda.

With all the thousands of things that can go wrong with the human body, causing hundreds of different diseases, it seems ridiculous to suggest that health and illness can be explained by a simple mechanism. But it's true. Here is the master equation: **Good things in, bad things out.** We are designed to take in good things, and any bad things that our bodies can't use (including the byproducts of the things we *do* use) are excreted. That is the body's natural instinct to detoxify. This process is present down to the cellular level:

Naturally, if bad things go in (junk food, toxic drugs, environmental toxins, etc.), there is more for the body to try to eliminate. But if not enough *good* things go in (vitamins, minerals, protein, fats, carbohydrates, enzymes, oxygen, etc.), things break down because of a lack of the elements the body needs to thrive.

Now if enough bad stuff does not come out (poor bowel movements, not enough exercise, not enough water, etc.), your toxic load is too great and you are poisoned from within. And things get really serious when good stuff comes *out*—as in cases of Irritable Bowel Syndrome and other conditions where food is expelled before the nutrients can be absorbed.

Health=good things in, bad things out. True at the organic level:

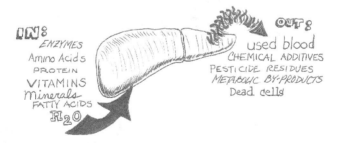

And true even at the mental/emotional level, as well:

FOCAL INFECTIONS

Another factor creates blockages to automatic healing in the body. That factor is focal infection. There can be tiny pockets of ongoing infection, or deposits of toxic substances, in odd places that go on for years. They can exist in areas far from the problem area and therefore never be suspected as causing or contributing to the chronic illness. They commonly occur in the tooth sockets, in the tonsils (or the tissue where the tonsils used to be), in the sinuses, or in the abdominal cavity. Focal infections are often present in the joints, in cases of arthritis[*].

Focal infections are often in locations that are not reached by normal drainage and thus escape the efforts of the body to automatically detoxify. They can also escape the effects of systemic antibiotic treatment. Since their presence is a constant irritant, the body produces inflammation in an attempt to burn them out and increase blood circulation to carry them off. When it can't do this completely, the whole process starts up again and you go through another round of the process. This is why some chronic inflammatory problems seem to get better every once in a while ("remissions") but always come back.

The organisms can travel to other parts of the body and create problems that cannot be recognized as being connected to the original site. Often, well-chosen treatments, nutritional supplements, and even drugs are ineffective because of this little-understood factor.

Rooting out focal disorders is an important part of turning around chronic illness. We use many ways to check for them, including skin conductance measurement, electro-

---

[*] Are you a physician? If so, do you think this idea is "alternative medicine" nonsense? Review *Cecil's Textbook of Medicine* (Saunders), Chap. 21. Also:

- Cucurull E, Espinoza LR: Gonococcal Arthritis. *Rheum Dis Clin North Am, 24:305, 1998.* An in-depth review of diagnostic and therapeutic modalities of disseminated gonococcal infection.
- Espinoza LR: Infectious Arthritis. *Rheum Dis Clin North Am 24:287, 1998.* An up-to-date review of the most common infectious disorders affecting the musculoskeletal system.
- Phillips, PE: Evidence Implications Infectious Agents In Rheumatoid Arthritis And Juvenile Rheumatoid Arthritis. *Clin EXD Rheumatol* 1988 6:87-94.
- Rook, GAW, et al: A Reppraisal Of The Evidence That Rheumatoid Arthritis And Several Other Idiopathic Diseases Are Slow Bacterial Infections. *Ann Rheum Dis* 52:S30-S38; 1993.
- Shetty AK, Gedalia A: Septic Arthritis In Children. *Rheum Dis Clin North Am 24:305, 1998.* A comprehensive review of septic arthritis.
- Tan PLJ, Skinner MA: The Microbial Cause Of Rheumatoid Arthritis: Time To Dump Koch'S Postulates. *J Rheumatol* 19:1170-71. 1992.

dermal screening, and thermography. Finding these blocks to healing and removing them can be the step that was missing in your health care up until now.

DENTAL HEALTH

Focal problems in the mouth are among the most common. Dentistry is a vital part of your overall health overhaul. Problems originating in the teeth or tooth sockets are behind a wide variety of chronic illnesses. This is because each tooth has bioelectrical connections to other parts of the body. As European medicine found back in the 1960s, a dental focus can cause disturbances down the line, just as sticking an acupuncture needle in the foot can influence the eye. There is a complicated network of connections stemming from the mouth, and it can be influenced by a number of common dental problems.

Silver amalgam (silver mixed with mercury) **fillings** have been in use for over a hundred years, and the official stance taken by the American Dental Association is that mercury does not leach out into the mouth in any significant amount. Don't believe this. There have been many, many studies since the 1920s, and even earlier, that show that this does in fact happen. Chewing, drinking hot liquids, and

DENTAL DANGERS
• Amalgam fillings
• Root canals
• Cavitations
• Impacted wisdom teeth
• Periodontal disease
• Mixed metals

changes in mouth pH all cause traces of mercury to be released into the body. Mercury is the second most toxic element known, a known neurotoxin that can cross the blood-brain barrier and thus affect the brain and nerve tissues. It crosses the placental barrier and can be transmitted to infants. It also has an affinity for the kidneys, depositing there easily. Mercury affects the white blood cells and reduces their ability to clear waste products. In this way, it is an immuno-suppressant. Dental mercury has been linked to dozens of diseases. We consider the replacing of amalgam fillings with non-toxic composite fillings to be one of the best investments you can make in your health. You must consult with a good, biologically-oriented dentist who follows the careful protocol for safe removal.

Root canals are another common problem that can have wide-ranging effects on a person's overall health. A series of experiments were done in the 1930s[*] where root canal-treated teeth were placed under the skin of laboratory animals. Nearly all the animals developed the same diseases that the donors (humans) had. Research showed that all root canal teeth are infected, regardless of whether they show any symptoms.

---

[*] Dental researcher Dr. Weston Price found links between root canals and heart problems; arthritis; kidney, liver, ovary and testicle infections; intestinal problems; skin conditions, and more. His research spanned 25 years, was published in twenty-five scientific papers and a two-volume textbook. This information was lost until George Meinig, D.D.S., wrote *Root Canal Cover-up Exposed* (Bion Pub.Co.) in 1993.

The bacterial strains and their proteins create a very toxic condition that will spread systemically. The immune system keeps this low-level infection at bay, until a greater load comes along—stressful situations, a viral illness like the flu, or some physical trauma—and then the tooth infection wins out. It typically strikes the most vulnerable area, which may be far from the mouth. This makes the connection between a tooth and a degenerative disease hard to spot.

Cavitations are the areas left after tooth extractions, where incomplete healing allows localized problems to occur that can affect the whole body. Poorly healed bone, bacteria that become sealed in the healed-over socket, or toxic material from the former diseased tooth can all create a systemic problem. A dentist, cleaning out these cavitations using a special dental tool and suctioning out the toxic debris, could be a major player in your journey back to health.

Impacted wisdom teeth have been found to be a factor in some bowel conditions, emotional problems, and neurological diseases like MS. Even when no infection can be detected, impacted third molars can cause problems and we are advised to have them cleanly removed.

Periodontal and gum disease can obviously cause wide-ranging problems as infection in the jawbone can travel through the whole skeletal system. The gums can be an outward signal that things are not what they should be, but they can also be an ongoing source of re-infection that keeps a destructive process going.

Mixed metals, in the form of gold fillings opposite silver fillings, will cause bioelectrical disturbances in the mouth that can affect the entire body. Let's say you have a silver filling in a lower molar, and you have a gold filling in the upper molar above it. Dissimilar metals when opposed tend to ionize, especially when the pH of the mouth becomes more acidic. They emit a tiny electric charge, literally making your mouth a battery. This impacts on your health over time. An early patient to our clinic had terrible migraine headaches that failed to respond to any treatment. When she came to us, she mentioned that she sometimes heard "beeping" sounds that no one else could hear. She said they sounded like Morse code. She was afraid she was "going crazy". An examination of her mouth revealed a row of gold-filled teeth on the bottom, and several silver amalgam fillings on top. We measured electrical conductivity at a number of points on the hands and feet, and the values varied drastically between when she had her mouth open and when she bit down. She was apparently picking up short wave broadcasts! Having the silver fillings replaced eliminated both the "beeping" and the migraines.

MENTAL HEALTH

Dental health and mental health don't seem like they have that much to do with each other (except rhyming). But they together illustrate one of the biggest goofs in conventional medicine: the mistaken idea that everything doesn't fit together. If you separate mental health from physical health, calling it (as they do) "Behavioral Medicine", it suggests that mind and body are not connected and need to be treated separately. Seeing dentistry as a separate field creates the illusion that, for example, cavities occur because of something happening in the mouth, instead of systemically in the body.

Just as there can be focal disturbances in the body tissues that create problems elsewhere, there are mental "foci" that create problems as well. Old trauma, unresolved issues, inhibitions, undeveloped sides of our personalities, injuries to our self-esteem, even completely forgotten events—all have an impact on our total health. **Anything that affects us emotionally affects us physically, and vice-versa.**

The traditional way to resolve these mental/emotional hurts is to talk about them and analyze them with a trained professional, hopefully improving the way we feel about them and fixing up life situations. This typically takes a long time. A more modern approach is to take a drug that re-arranges brain chemicals so that you feel better even if you haven't fixed up the problems. This, as you know, is very popular. Of course, if the drug is not taken, the problems are felt again. But the physical problems that result from the unresolved conflicts don't go away just because someone is taking an anti-depressant. No one notices this connection because it seems to be a different problem. You go to your family doctor for your physical treatment and you go to your therapist for your mental health. No one is looking for a mental cause for the physical illnesses that are progressing.

Happily, there are breakthrough discoveries about how the mind-body connection works, that make long, drawn-out psychotherapy obsolete. And, because they can work rapidly, they are serious competition for drug therapy. Don't expect to hear about these new ideas from the field of "Behavioral Medicine" that relies on psychotherapy and drugs!

There are a number of new treatments with different names and slightly different approaches that use mechanisms that have never been well understood in the field of psychology. Neuro-Linguistic Programming, EMDR (Eye Movement Desensitization and Reprocessing), Thought Field Therapy, Neuro-Emotional Therapy, Psycho-Kinesiology, Transformational Kinesiology, and Emotional Freedom Technique are just some of the various methods[*].

The basic idea is this: Negative emotions are not the direct result of a traumatic event. Nor does a person's thoughts about the event or issue create the problem directly. What happens is that disturbances in the mind-body system (what some call the "Thought Field") create changes in nervous system activity, hormone and other chemical levels, and changes in the bioenergetic pathways that link the organs. Simply put, this is encoded information from the mind that triggers a host of very tiny responses. A shifting of brain chemicals can cause depression, but it can also cause changes in organic functioning. It is the information itself that causes the shift in

---

[*] There are also some methods that use the same kind of mechanism for the treatment of autoimmune illnesses, like NAET (Nambudripad Allergy Elimination Technique) and JMT (Jaffe-Mallor Technique).

33

chemical balance, and that information can be erased. When the information is erased, how you feel and think about the original cause is immediately changed.

Erasing the informational triggers involves tracing the tiny electrical disturbances they cause. Then, by changing the bioelectrical flow through that pathway while the person focuses on the problem, it renders the disturbance in the person's mind-body system inactive. **The change can sometimes take place in just minutes.** The process is comfortable, painless, and the effects are immediate and long lasting. Even complicated cases can sometimes require only a few sessions—far, far fewer than in conventional counseling or therapy.

There are certain types of mind-body "blocks" that are associated with negative thinking and self-defeating behavior, and these are known to be a major factor in why some people do not heal (psychologically or physically) in spite of intensive treatment. Now there is a way to correct this situation so that medical or psychological treatment becomes effective again and is not needed as much.

Although this type of therapy, like any other, does not work for every problem in very person, studies show that it is over 90% effective for the most common problems, such as emotional trauma, self-esteem problems, depression, anxiety, grief, and addictive tendencies. Since you may be wondering, the person does not have to believe in the therapy in order for it to be effective. In fact, research has shown that even the *therapist* does not have to believe it will work. Many mental health professionals balk at the suggestion that something so simple can work so quickly and create such a dramatic emotional transformation. Skepticism disappears when they experience these techniques for themselves.

Homeopathic medicines, which will be discussed in more detail later, are also of great value in problems of a mind-body nature. Because the highly diluted substances used in homeopathic medicines gain in subtle power as they are processed, they can be dramatically effective in mental and emotional states when given in their higher potencies. The Homeopathic doctor will painstakingly match up your individual characteristics and the symptoms of your problems with a medicine that fits as perfectly as possible. When it really fits your "keyhole", it can "unlock" the barrier to your getting well. I have seen depression, phobias, compulsions, and even autism change for the better with only the administration of just a few doses of a highly diluted remedy. It's not as simple a process as it sounds, but when it's done right, it's like magic.

34

SOME INTERESTING FACTS TO CONSIDER

Each year, the average American adult consumes: • 5 pounds of potato chips • 7 pounds of corn chips, pretzels, and popcorn • 18 pounds of candy • 20 gallons of ice cream • 50 pounds of cakes & cookies • 55 pounds of fats & oils • 63 dozen donuts • 100 pounds of refined sugar • 200 sticks of chewing gum • 300 cans of soda pop	• 7% of adults are active alcoholics (one in fourteen) • 36% of adults still smoke (one in three) • The average American will live to age 76.... the average M.D. lives to age 68. • The U.S. spends $20 million *per minute* on "healthcare"; more per capita than any country on the planet. • Over 50% of all healthcare costs are spent keeping people alive in *the last 5 days* of their lives.
In 1900 the risk of cancer was 1 person in 30. In 1980 the risk of cancer was 1 person in 5. In 1990 the risk of cancer was 1 person in 4. In 1995 the risk of cancer was 1 person in 3. In 2000 the risk of cancer was 1 person in 2.	How accurate are the standard hospital diagnostic tests? The error rate is about 20%--and half of the mistakes lead to death. *Source: Washington University in St. Louis Feature Service, September 1985*

Conventional medical treatment is now the leading cause of death in the United States.

A total of 783,936 deaths per year from adverse drug reactions, unnecessary procedures, medication errors and other hospital errors, and hospital-acquired infections.
Death from heart disease is second at 699,697. Cancer is third at 553,251. Reactions to *properly taken* prescription drugs (305,000 deaths per year) itself is the fourth leading cause of death). In addition, 1.5 million people require hospitalization for treatment of side effects per year (36% of all hospital admissions).

Sources:
Nutrition Institute of America report, November 2003
BMC Nephrol., December 22, 2003
Am J Med, August 1, 2000
JAMA, July 30, 2000
Center for Health Policy Research, George Washington University Medical Center, Washington, D.C. 1999 (Thomas J. Moore, et. al.)
National Institutes of Health, 1999

EVALUATING YOUR INDIVIDUAL HEALTH NEEDS

PHYSICAL EXAMINATION
- We make use of the usual methods of assessing bodily functions such as listening to the lungs and heart, pulse and tongue analysis, etc., and palpation of affected areas. Attention is made as to how the spine and extremities line up, and how your body moves through space and against resistance.

OXYGENATION TEST
- Evaluation of the body's oxygen-carrying ability gives an indication of how you will respond to treatment, and is a good indicator of progress, when re-checked.

BOWEL TRANSIT TIME
- This "low-tech" test gives an accurate estimate of how efficiently you are eliminating waste.

ABI TEST
- This simple, non-invasive test gives an estimate of the degree of atherosclerosis and poor circulation present.

MI FACTOR TEST
- A formula, using the relationship between your cholesterol readings and your blood pressure, reveals your chances of heart attack or stroke.

HEART RATE RECOVERY TEST
- This test assesses quickly your heart rate returns to normal after exercise and is useful in determining cardiovascular health.

BASAL BODY TEMPERATURE TEST
- The body's core temperature is a reflection of how well your thyroid hormones are performing, which have an influence on body metabolism, enzyme levels, and cardiovascular health.

ELECTRO-DERMAL SCREENING
- Electronic assessment of the acupuncture points and meridians, as is done in the most modern Functional Medicine hospitals and clinics. Conductance readings taken on test points on the fingers and toes reveal the locations of electrical imbalances inside the body. This European diagnostic method is also used to test for traces of low-level contaminants, residuals of old infections, or food sensitivities, which might contribute to ill-health.

INFRARED THERMOGRAPHY
- This safe form of diagnostic imaging uses no ionizing radiation as do X-rays, CT scans, or mammography. Infrared thermography can be used to detect the locations of internal problems by making visible the changes in blood flow at the site. Improvement due to treatment can be tracked with this method.
- Routine breast cancer screening can be done with the same technology. This is recommended to all women as a reliable early screening device (mammograms

36

are actually a late form of visualizing tumor formation). This complements mammography and reduces exposure to unnecessary radiation.

SALIVA TESTS
- **Alkaline Buffer Challenge**—An excellent screening for mineral deficiency. This saliva pH test helps us to detect deep levels of physical and mental stress.
- **Zinc Test**—Many patients today show signs of Zinc insufficiency. This simple test will instantly determine your Zinc status.

URINE TESTS
- **Adrenal Stress (Koenigsberg) Urine Test**—This is a specialized urine test that measures the amount of chloride displaced into your urine. Excess chloride in the urine is a reliable measurement of adrenal stress or adrenal fatigue.
- **Vitamin C Test**—accurately determines your Vitamin C status; how well your body is holding onto its supplies of Vitamin C.
- **Calcium (Sulkowich) Test**—With this urine test we can accurately determine whether your calcium level is adequate, low, or even too high.
- **Oxidata™ Urine Test**—50 times more accurate than blood serum free radical tests. Free radical damage can lead to cell degeneration, initiating a host of diseases such as fatigue, allergies, arthritis, lupus, elevated cholesterol, and degenerative heart disease.
- **Malabsorption (Obermeyer) Test**—Detects the presence of harmful anaerobic bacteria and bowel dysbiosis. This can lead to malabsorption, digestive disturbances, allergies, and inflammatory symptoms.
- **Chem Panel Urine Test**—Measures 13 separate urine categories. We use this test to screen you for serious health problems including infections, tissue degeneration, liver disease, kidney disease, and diabetes.
- **Urinary Bone Marker Test**—Bone density and osteoporosis can now be evaluated with a simple lab test.
- **Hormone Panels**—Male and female hormones can be assessed through the urine and saliva instead of the usual blood tests, which measure protein-bound hormones and do not give an accurate picture.
- **Total Sugars Test**—measures total sugars, not just glucose. Identifies whether you are a glucogenic or a ketogenic type and how to address it.
- **Nitric Oxide levels, Proteins, Urea, Nitrates, Ammonia, and Electrolytes**—all can be tested to evaluate protein utilization, fluid transport, energy production, liver and kidney function, and pH adaptability.

BLOODWORK
- All appropriate blood tests can be ordered if needed to arrive at a more conclusive evaluation of either your original state or as a confirmation of improvement after treatment.

STOOL ANALYSIS
- Simple home checks for hidden blood are routinely done. More involved tests are used to check for parasites and other conditions.

HAIR MINERAL ANALYSIS

- FDA-approved hair analysis tests for mineral content and for measuring amounts of heavy metals in the body can be done to document problems and measure improvement.

SOME OF THE TREATMENT METHODS USED AT INFINITY

ACUPUNCTURE— Acupuncture is approved for over 150 different conditions by the World Health Organization, and it matches well with our other types of treatment. We have been making use of acupuncture in the clinic since 1978. We use painlessly thin disposable Japanese needles.

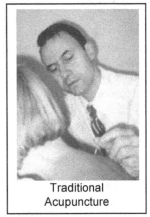

Traditional Acupuncture

There are a number of ways to stimulate acupuncture points. Many of these were developed in Europe over the last fifty years, but acupuncture as a medical tool in the United States tends to favor the older Chinese approach. For that reason, you may not have heard of these other methods. Although most people associate acupuncture with the Far East, the medical communities of France, Germany, and Russia have developed it to a high degree.

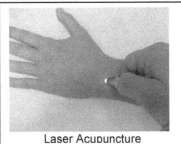

Laser Acupuncture

I more often use non-needle approaches to acupuncture, including stimulation with a mild electrical current, soft lasers, or a combination of a pulsed magnetic field coupled with a long wave infra-red beam. Although the person feels nothing, this type of stimulation discharges an acupuncture point far faster than a needle. Those who don't want needle acupuncture can be treated effectively with this method without needles. Even small children enjoy it.

I also use a novel method of stimulating acupuncture points with even greater effect. First, I choose the specific points to be treated. Instead of inserting needles, I place a small dab of liquid herbal/mineral formula (chosen according to your condition) onto the skin over the points. Then a mild electric current is used to stimulate the skin. This current is carried into the points more strongly because the formula conducts electricity. But the electric current helps drive in the ingredients, creating a kind of "injection" of the herbal/mineral formula without breaking the skin. This is called *iontophoresis*, or ion transfer.

Non-needle acupuncture

It is common among some practitioners to inject medicinal substances into the acupuncture points to stimulate them. Our unique method has the benefits of both traditional acupuncture and point-injection, without using any kind of needle.

MERIDIAN BALANCING TEST AND TREATMENT—This modern version of traditional acupuncturists' "Seasonal Balancing" treatment takes electrical readings from the various energetic pathways (meridians) in the body that lead to the internal organs.

Using a special probe, electrical conductivity is measured at test points that relate to the major systems of the body. Readings on a 0-100 scale are taken, and the abnormal highs and lows are noted. By reducing the charge in the ones that are overactive, and increasing the charge in the ones that are deficient, a meridian balancing occurs. Our long-time patients come at each change of season for this treatment and all agree that it keeps them healthier through the coming months. Excellent for spotting early signs of dysfunction, before symptoms develop.

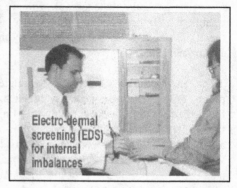

Electro-dermal screening (EDS) for internal imbalances

METABOLIC BALANCING— Restoring balance to the body's biochemical state, or the **biological terrain**. Correcting such factors as acid-alkaline balance, maldigestion, malabsorption, cell oxidation or reduction, and vitamin or mineral deficiencies will normalize the biological terrain, which is the foundation for building up health.

NUTRITIONAL SUPPLEMENTATION— Using high-quality supplements to correct deficiencies of essential elements in the body is something we do not neglect. This idea is nothing new. Once, it was only the "health nut" who took vitamins. Today it is considered the smart thing to do. However, since vitamin tablets have become commonplace and can be purchased through many different outlets, people are taking products that do not pay off in the ways they expect. For one reason, they may have difficulty absorbing the nutrients because of biochemical problems, as we mentioned in the previous section on metabolic balancing.

Another reason is that, although they may be buying a "specially formulated" vitamin product by their favorite featured-on-TV doctor, they are likely getting an overpriced version of what they could buy off their local store's shelf. There is the perception that expense means quality, but the reality is that all the dozens and dozens of companies making such products get their raw materials from only a small number of sources. That's right—the expensive vitamins *and* the cheap vitamins in the USA are made from the materials coming from only a few (are you ready for it?)...*chemical companies*. Not places called "Pure Land Farms", "Earth Natural Co-op", or "Blue Skies Encapsulations". Chemical companies. Why? Get ready for the next shock.

You may never have seen a real vitamin. Although the label on the bottle you picked up at the health food store says "natural", it likely is not. What is it if not natural?

Back at the beginning of the nutritional supplement industry, manufacturers found that the cheapest, most profitable way to make vitamin tablets was from *synthetic analogues* (instead of the real vitamin found in food), *isolates* (one part of the vitamin complex plucked out of its naturally-occurring group of substances), and *minerals made from rocks* (instead of the easily-assimilated forms found in food). Making tablets from natural vitamins and minerals involved more stages of production, required lots of farmland with excellent soil, and created tablets that did not have a long shelf life and were sometimes hard to swallow because of the size.

Synthetic vitamins are not as easily absorbed and utilized. But there is another problem. Taking synthetic analogues can block your body's ability to absorb the *real* nutrient from your food. Taking an isolated nutrient instead of the whole vitamin or mineral complex can cause a deficiency of the other nutrients in that group. And when I say, "minerals made from rocks", you are probably wondering, "Where *else* would they come from?" Well, the minerals you absorb easily and rely on for your health come from food. If you suck on a rock you will not absorb much in the way of minerals. Manufacturing companies use a solvent to leach minerals out of rocks very cheaply, in order to get those minerals into a form that can go into a tablet (along with the residue of the solvent).

So when you are taking magnesium gluconate, for example, it is a rock treated with gluconic acid to access the magnesium. This is why you are advised not to take magnesium gluconate when you are pregnant or nursing—the gluconic acid is not known to be safe for the baby. Rule of thumb: If your mineral has a two-word name, it is not in its natural state.

And the minerals you don't absorb? They join together to create mineral salts (which your body can't use) and tend to deposit. Where? In the connective tissues, for one. When someone with arthritis gets that sharp pain from moving a joint, I know that there are mineral crystals digging into the sensitive nerves there. I know because it has happened to me. Even scarier, though, is the deposition of minerals in the arteries. This is how plaque gradually clogs up your blood vessels.

Minerals in food are different. The magnesium in bran or buckwheat is just that— magnesium. Blueberries and green peppers are rich in the Vitamin C complex—not just ascorbic acid, which is *one* of the two dozen components contained there. Are you starting to see the difference? A couple generations have now come to believe that ascorbic acid *is* Vitamin C, instead of something that is *in* Vitamin C.

As you probably know, the Food and Drug Administration is the Federal agency that has the responsibility of overseeing all products taken for health. When it was called upon to regulate vitamins, it adopted the industry standard. When determining how much of a nutrient the average person needed to consume daily, the substance under study was the synthetic or isolated version, not the naturally found form.

41

This is the USP (United States Pharmacopoeia)* standard. For this reason, when the "Daily Value" (DV) appears on your vitamin bottle label (as it must, by law), the calculated need is an amount that is not necessarily accurate for natural nutrients. It may take 1200 milligrams of calcium in the form of calcium citrate to prevent bone loss, but likely far less of the active, natural calcium from spinach or almonds. Of course, you can take in more calcium more easily by swallowing tablets made from concentrated spinach or almonds, and that is exactly the advantage in taking natural vitamins. Also, even if you are eating those foods regularly, you have no assurance that the soil they were grown in had much mineral content (it likely didn't).

It may be occurring to you now to switch to a different brand of vitamin supplement. I have one more disturbing bit of news. The small number of manufacturers who make truly natural vitamins all distribute their products mostly through doctors. There is a tremendous prejudice against these products in the industry. Retail stores will not stock these products. Periodicals do not even accept their advertising! Even though I was trained in the field of natural medicine, I might not have known this if my own relative had not founded one of those companies. The FDA had him put in prison, back in the 1950s. His crime? He would not be quiet about the scientific studies showing the impact of food-based nutrients on disease. He was also attracting a lot of attention to the fact that public school lunch programs were nutritionally deficient. This was radical behavior in the '50s.

He was not the only one who was harassed. All the founders of natural nutrition firms underwent great persecution. Because of this situation, the companies making natural supplements decided long ago to market to doctors and not the public. When licensed physicians were presented with the facts and they prescribed the products based on their professional judgment, the Federal regulating agencies had a harder time stopping the sharing of nutritional information and access to natural vitamins and minerals.

And so it is that we use these natural supplements made from concentrated organic foods in our clinic. I never wanted to be in the business of having to stock such products in our dispensary, but it was the only way for my patients to have the products they needed. We have people tell us frequently how different they feel taking the natural products compared to the typical USP vitamins they were taking before they came to us.

---

* This is the giant book that contains the approved formulations, dosages, and standards of manufacturing for all prescription medicines, over-the-counter drugs, and other health products. It is overseen by the American Medical Association, The American Pharmaceutical Association, and the Food and Drug Administration.

MANUAL THERAPIES— Methods such as pressure on the acupuncture points, massage, joint mobilizations, soft-tissue manipulations such as myofascial release, and re-education of neuromuscular functions by positioning and guided movement, may be used for problems of a musculoskeletal nature. One of the interesting things about the field of natural medicine is how different providers perform manual therapies. For example, many people are familiar with the manipulations of a Doctor of Chiropractic. Those who have been treated by an Osteopathic doctor (specifically an Osteopath who uses manipulation) will say that it's slightly different than the Chiropractor's maneuvers. Naturopathic doctors have yet a different way of restoring motion to the joints. And Doctors of Oriental medicine have been using such methods for the longest time of all of these practitioners. Each group has a unique way of seeing the problem and a different theory of relieving it. At our clinic, we use methods that correct

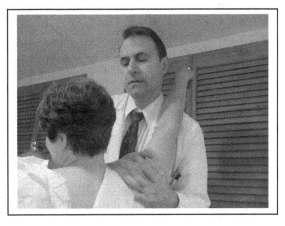

these imbalances without a lot of forceful thrusting, and patients tell us that they prefer it to other types of joint correction.

OTHER THERAPIES— Laser stimulation, ion pumps, magnets, controlled-spectrum light therapy, infrared and ultraviolet, diathermy and ultrasound, and hydrotherapy are all used in accordance with Naturopathic and Oriental medical theories. Let's look at some of these:

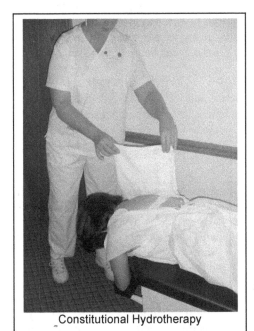

Constitutional Hydrotherapy

Hydrotherapy
There are many ways to use water for healing. Today, only a few are used in standard physical therapy. For that reason, most people only think of a whirlpool bath or swimming exercises when you say "hydrotherapy". Actually, water can be such a potent force that both local and general conditions in the body can be changed very rapidly by the precise application of hot and cold water to the body's surface. One method we use, called **Constitutional Hydrotherapy**, has been in use since the 1930s and is

really quite amazing in the way it benefits chronic conditions and boosts resistance to illness. If water is so powerful, why is it not used more often as a treatment in hospitals today? Just picture drug companies trying to patent terry cloth towels and water!

In the old natural medicine sanatoriums, a service that was always present was the Russian Bath (also called Turkish Bath). Not truly a bath, but a way of using concentrated steam to purify the body, it differs from the Finnish *sauna* by its use of moist vapors instead of dry heat. While it can be applied in full rooms like saunas, it has traditionally been applied for healing in the form of a steam cabinet. A person can remain in the cabinet far longer than he could in a steam room, because the head is excluded, and normal room temperature air can be breathed.

Russian Bath

But the temperature inside the cabinet—that's another story! Steam cabinets are a way of recreating the healing action of a fever. Tremendous amounts of impurities are released from the body through the pores during a session. The soothing heat penetrates the tissues, and the general metabolism is stirred to greater activity. This is even more so when we combine the potent effects of extra oxygen in the steam cabinet, and various therapeutic plant extracts and oils can be used at the same time to maximize effects.

Detoxification can be achieved by many means, and we use them all. But the steam cabinet is one of the most powerful and fast acting. A short time spent there gives a great boost to overall health and is specifically needed for many conditions.

After hours, members of our staff climb in for a quick steam also! It's just the thing when you are dealing with sick people all day long, and want to keep up your elimination of toxins.

Another form of treatment, which is almost forgotten today, is called the Electro-Galvanic Bath. This is the use

Electro-Galvanic Bath

44

of pure water with the addition of various mineral salts, and sometimes herbal extracts, used to bathe different body parts. But the difference is this: a mild electric current is carried through the water. Sound dangerous? It isn't. The reason for the electricity is that certain frequencies have their own therapeutic effects in the body, but when the polarity of the electricity is right for the condition under treatment, it creates a pulling effect on substances you want to expel from the body. Moreover, it can be used to drive the medicaments in the water into the tissues. Once again, this creates a kind of "injection" without needles. It is a wonderful way to get certain herbal and mineral compounds into places like underneath the finger- and toenails, into the soft tissue around painful joints, etc.

Electro-Galvanic Bath was used for many, many years in physical therapy clinics but has largely disappeared over the years. This is an example of a valuable tool that we will not allow to die.

Phototherapy

You might think this means "feeling better by looking at photographs". No. It is a more technical term for "light therapy". There are many portions of the light spectrum that can be used to stimulate healing.

Infrared rays (both near and far infrared) have a well-established use in treating a variety of health problems, and can be used for detoxification as well as to reduce inflammation in painful joints.

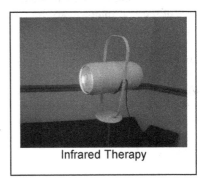
Infrared Therapy

Ultraviolet rays, of course, are often used to treat skin conditions, but they create many positive chemical reactions in the body, including the activation of Vitamin D and increasing the uptake of calcium to the bones—something very beneficial to those with osteoporosis or osteopenia. Our clinic uses a device that produces the full range of ultraviolet rays; it is in fact artificial sunlight. And by now most everyone knows that sunlight or full-spectrum light is prescribed for those suffering from seasonal depression.

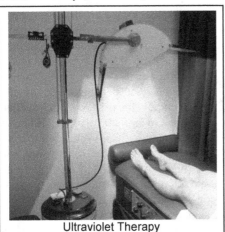
Ultraviolet Therapy

Chromotherapy

The most extraordinary use of light that I have seen, though, is the use of the visible spectrum—in the form of colored light. Years ago, much research was done to verify that certain colors had an impact on certain conditions and tissues in the body. As more and more medical doctors began to use this therapy, the growing pharmaceutical industry stepped in and eliminated it. Almost. The lucky ones of my generation, like myself, had teachers from the day when this therapy was taught.

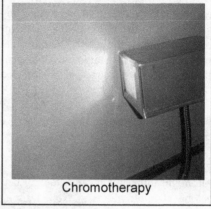
Chromotherapy

My mentor was ninety-one years old when I met him, and he had studied with three of the four major innovators in color and light therapy, back in the 1930s. I was very fortunate to have learned from him over six years the secrets of treating with various frequencies of light.

We find that when the appropriate color of light is applied, acupuncture and other therapies work faster and better. In fact, there is a way of using colored light instead of needles to do acupuncture. Again, colored light is cheap and effective, so here is another valuable healing agent you have never heard about in the era of big money medicine.

Ultrasound

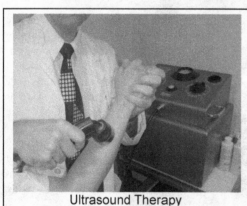
Ultrasound Therapy

Almost everyone has heard of ultrasound, or maybe has had it done by a physical therapist at one time. This is not diagnostic ultrasound, where an image of the inside of the body is produced. Ultrasound *therapy* uses sound waves that penetrate and rapidly pummel the tissues in a way that reduces inflammation and breaks down deposits. Although you don't really feel much of anything, the high-speed sound waves create an internal heat inside as the therapist rubs the sound head over the surface of the skin. An interesting application of this is when a medicated substance is placed on the skin, and the waves can drive it into the tissues, a process called *phonophoresis*. This is just as I mentioned before when discussing my method of using electrical stimulation on the acupuncture points to "inject" a herbal/mineral solution without needles. Sound waves will cause a transfer of these formulas also.

46

Short Wave Diathermy

When I was learning natural medicine, this was a therapy that would not go away. Three of my preceptors during my training years used this type of treatment. Diathermy, along with ultrasound, has always been a prominent tool of the physical therapist as well as the natural doctor. In recent years, it has been the tendency of physical therapists to forget about diathermy and rely on ultrasound and other newer methods. This is a shame because it is so valuable.

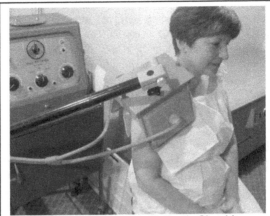

Diathermy Applied to a Patient's Shoulder

Short wave diathermy uses, as the name implies, radio waves. Like ultrasound uses sound, diathermy uses radio waves to penetrate the tissues and creates a soothing internal heat that reduces inflammation and congestion, rapidly pumps waste products out of the connective tissues and fluids, and draws new

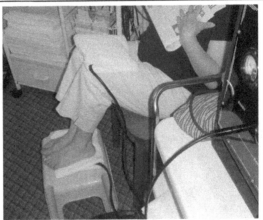

Diathermy Applied Through the Feet and Legs to Help Circulation

blood and nutrients into the area. In this way, it speeds the rate of healing. Diathermy can be used on pains, sprains, and strains, but also on bronchial congestion, sinusitis, and similar conditions, because radio waves pass through tissue and can get deep into the "nooks and crannies" that other therapies can't reach. Its outstanding success in relieving sinus pressure alone has been a blessing, because we live in a region where a high percentage of the populace has sinus problems.

NATURAL MEDICINES

"Natural medicines" takes in a lot of territory. There are several categories that qualify for this description. Herbal, mineral, or organic extracts all fit this title. Many traditional remedies in all schools of medicine come from these sources. Since the most common type of natural medicine used is plant or herbal, we will discuss this first.

Botanical Medicines

Many patients seen in our clinic today are already taking herbal medicines. There is a general perception that "herbal" equates with "safe", because it seems like a return to a simpler time when Grandma would mix up poultices from plants found out back and treat the kids. Herbal medicines can be powerful, however; many of them interact poorly with prescription drugs the patient is already taking. A well-trained prescriber of botanical medicines (as they are referred to by the profession) will always cross-reference the actions of a particular plant with the actions of a drug being taken to avoid any accidents.

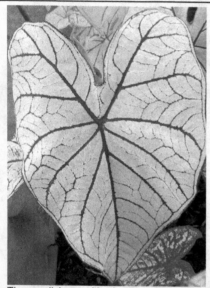

The medicines with the broadest-acting constituents are those derived from plants. Thousands of these factors have been identified and their effects on diseases have been rigorously tested. Anyone who tries to tell you that plant-based medicines are "quackery" simply doesn't know the scientific literature—or he's lying.

At one time most drugs listed in the United States Pharmacopoeia (USP) were plant medicines. Over the years, drug companies gradually eliminated most of the pure botanicals from the official formulary, because synthetic drugs are far more profitable—*not because the herbs didn't work*. Even so, about 30% of the synthetic drugs on the market today are derived from plants in the first place. Naturally occurring chemicals in the plants are altered in a way that allows them to be patented, because they are new compounds. Pharmaceutical companies profit by having everyone think that "real" medicines are created out of nothing with chemicals. This is not accurate. I will discuss this in more depth shortly, in *Making Sense of Herbal Medicines.*

Since many people are rebelling against the common synthetic drugs and their side effects, it is understandable that they would turn to herbal medicines. Yet, the items that they have access to may be less than effective. First of all, the herbal teas that they buy from the local health food store or over the Internet may be nothing more than the "grass clippings" left in the bottom of the barrel after other companies have purchased the larger portions of the fresh and vital herbs. Companies manufacturing

48

to pharmaceutical specifications have more money to spend and acquire larger amounts of the wild crafted or organically raised herbs. So the potency of herbal medicines can vary widely, not only from company to company, but from product to product within a single company.

Another problem has arisen in recent years. The major companies supplying the natural medicine profession have begun concentrating on isolating specific, so-called "active" ingredients in plants and are insisting that we prescribe these instead of the whole plant, like Grandma used to. The promise is that the potency is standardized and its efficacy is reliable, but (as I will explain soon) this is not necessarily so. As a result, the earnest seeker of health either buys a product that is too weak or adulterated to have the effect he wants, or buys a supposedly superior product at great expense that has side effects just like the drug he tried to get away from!

Making Sense Of Herbal Medicines

There are "empirical" herbal experts who understand the healing properties of plants, and that is where Grandma got her knowledge. Some train in formal programs and are granted a degree, such as Master Herbalist (M.H.).

Of course, plant medicines have been used for hundreds of years in Oriental medicine. Because the laws may differ depending on the state in which you live, a provider may only be able to administer acupuncture but not herbal medicines in some places. But this does not mean that it is not a part of the overall system.

Plant medicines have also been traditionally used in Naturopathic medicine. Before there was a large industry producing herbal medicines for both the natural doctor and the lay consumer, Naturopaths treated patients with botanical (herbal) medicines in various forms: fluid extracts, tinctures, infusions (teas), decoctions (boiled herbs), and the dried powdered herbs themselves. All the methods of preparation captured the healing properties of the plants and delivered them to the patient.

Another group of doctors, the Eclectic physicians, also used plant medicines as their main weapons against disease. The Eclectic medical profession developed the first standardized processing of herbal medicines. Today it is hard to believe, but all the medicines used by all Eclectic physicians came from *one* pharmaceutical company, Lloyd Brothers of Cincinnati.

As we mentioned before, all official drugs in this country are listed in a government publication called the United States Pharmacopoeia (USP). At a time when a high percentage of the drugs in the USP were herbal extracts, the Eclectics relied upon one firm to produce consistent quality botanical drugs, called (accurately enough) "Specific Medications". Two things were "specific" about them. First, that there was no question about the identity and potency of the chosen medicine. Gone were the days of someone gathering herbs in the woods to make teas for sick people and accidentally picking the wrong variety of plant. Herbal medicine was now a very scientific endeavor. Lloyd Brothers' laboratories insured that every bottle of a particular medicine had the same potency as all the others.

Second, they were "specific" because the predictable effects of a certain herb on the body never varied. They knew, for example, that the plant *Phytolacca decandra* always acted on the lymphatic system, while *Chelidonium majus* always acted on the liver. Different medicines had different actions in the body, depending on the acid/alkaline balance of the bodily fluids, and they were prescribed according to the patient's situation. Eclectic MDs used plant medicines in a very precise way, while regular MDs, who also used many plant-based medicines, prescribed larger doses in a "take-this-for-that" fashion.

By contrast, Eclectic doctors examined their patients for signs of internal dysfunction and used medicines that acted

LLOYD BROTHERS, PHARMACISTS INC.
CINCINNATI, OHIO

on the underlying conditions that resulted in the illness. They recognized that illnesses were made up of a collection of factors or conditions in the body. Rather than use a drug made for a specific disease, Eclectics used drugs for specific aspects of a sick person's case—not for the conventional diagnosis. The doses were just large enough to have a stimulating action on the targeted organ or function, but so small that they were not toxic. Eclectic treatment was very effective and without the typical side effects of the regular doctor's strong medicines.

Over time, drug companies became more powerful and produced more and more synthetic drugs. Eclectic medical schools faded away. Lloyd Brothers' Specific Medications were no longer available by 1960. Naturopathic doctors had no such firm to provide them with high-quality herbal medicines, and remained dependent on a loose network of herb providers. In modern times, there wasn't much credibility in a bag of twigs that the patient had to boil at home. This hampered the image of Naturopaths as prescribers of oral medicines, and they became known primarily as providers of physical therapies and lifestyle management. Gradually, new companies began to emerge that made reliable, easy to use, and respectable herbal products for the natural medicine profession.

But enough time had elapsed that the public lost sight of the fact that fully thirty percent of the drugs listed in the USP still came from plants. The average consumer believed that "real" medicines were made in test tubes, not grown. Health care providers who used botanical medicines were seen as quaint throwbacks to an earlier, less scientific time. After all, if herbs worked that well, they wouldn't have been replaced with synthetic drugs, right?

By the late 1970s, the trend started to reverse. Interest in natural foods and natural health care increased sharply. After licensing had dropped from over two dozen states down to eight states, Naturopathic doctors started returning. New schools were built. New awareness on the part of the public allowed companies producing vitamins and natural supplements to begin producing "nutriceuticals"—a whole new category of health care product.

Regular medicine did not take notice as quickly as the pharmaceutical industry. But soon both were decrying herbal products as outmoded, ineffective, and dangerous. In two ways this propaganda misled many otherwise intelligent people: One was the impression that herbs couldn't be that reliable as medicines, since they weren't regulated like synthetic drugs; and the other was that since they *were* unregulated, the public was not protected against dangerous products.

Our clinic pharmacy houses hundreds of plant-based medicines.

Anyone who studies the topic in depth will know that regulation of drugs began because of problems with the safety and efficacy of *synthetic* drugs, not natural ones. The requirement of a prescription was a measure that came about because of man-made, patented drugs. They were often totally ineffective for their advertised purposes. They also often contained addictive or otherwise harmful substances. The 1938 Food, Drug, and Cosmetic Act was made law after a public outrage over a Tennessee drug company's pediatric drug (containing a solvent similar to antifreeze) that resulted in the death of over 100 people—mostly children.

It was already established that herbs are safe enough and effective enough to be available to the consumer. This is why edible herbs (but having a medicinal effect) that are already present in the food supply are not regulated like drugs are. The FDA does require proper labeling on herbal products, however, as it does on all foods. *Drug regulation was never meant to show that drugs work and herbs don't, but to see that medicines are safe enough and valuable enough stay on the market.*

Now why would drug companies have been so anxious to introduce synthetic drugs? Answer: *They can patent them* (because they are not found in nature). A patent will give a company the exclusive right to market a drug, insuring that it will make a lot of money.

But why do the well-advertised, profitable, and public-trusted drugs all have side effects that the public hates? Read on.

Botanical medicines are used all over the world; they are not representative of jungle lands and third world countries. They are used in highly industrial countries that have advanced science but also have cultures old enough to respect the long history of successful use of natural medicines. More than half of physicians in most first world countries believe in using natural medicines. They also scoff at the current American tendency to take a beneficial herb and try to isolate the "active ingredient". This is a mentality, they say, that is an outgrowth of a culture that is governed by drug companies.

Unfortunately, with the rebirth of Naturopathic medicine came a new direction. Rather than succeed on their own merits and practice as their forerunners had, the new generation of Naturopaths has decided to become "respectable" by only using those products that have passed judgment by the mainstream scientific community. If a number of studies have been published in a "name" journal, if a laboratory has located the specific alkaloid in a plant that supposedly gives the medicinal effect, it is accepted. Botanicals have come to be viewed more and more like synthetic drugs.

Nutritional supplements and natural medicines of all kinds make their way from our dispensary to all parts of the country.

The idea of a beneficial non-toxic herb having its "active ingredient" isolated (as if there is only one) is laughable. The naturally occurring compounds in the plant that are discarded are beneficial as well. While the isolated ingredient may have a chemically stronger effect on its own, it usually has less of a restorative effect on the organs of the sick person, and greater side effects than if the whole herb was taken.

Once again, why would they be anxious to do that? An isolated, single ingredient can often be patented (perhaps with a molecular modification). A whole plant cannot be. But even more disturbing: In an age where the accountants determine the direction of a company, it is an ever-present concept that drugs with side effects will bring repeat business. Even a fractionated herb, which fails to completely cure, can create new problems that will require other herbs to relieve. Sound familiar? Remember what happens when you take an isolated vitamin? A deficiency of other vitamins in the complex. This realization is not lost on the industry, which is why more and more often drug companies are acquiring nutriceutical and botanical medicine companies.

Botanical medicines made from whole plants are more beneficial, and have fewer side effects, but are not as profitable. Isolated ingredients are often very effective for eliminating a specific symptom, but are less beneficial for correcting underlying problems. They have more side effects, but have much greater potential profit. One of the things that actually make them profitable is that they can suppress symptoms but might not fix anything! It is disappointing to traditional Naturopaths and other botanical prescribers to see that the vanguard of natural medicine in the 21st Century is simply a copy of the regular medical establishment.

Nutriceuticals

As mentioned, today there is a new class of natural medicine using nutritional substances for their medicinal effects. For example, Vitamin C can be used to fight infections, Vitamin A can be used for certain vision problems, fish oil benefits cardiovascular problems. Even though they may be necessary nutrients in the diet, they can be used to increase the body's healing action in certain areas. Called *nutriceuticals* (or *nutraceuticals*), they are often vitamins, minerals, amino acids, etc., combined with herbal extracts and formulated for specific kinds of conditions. Another rapidly growing type of product is the *cosmeceutical*, a nutritional product that is used for cosmetic purposes.

Homeopathic Medicines

While both Naturopathic doctors and Doctors of Oriental medicine typically use herbal medicines, there is one other area of specialty that not all practitioners share: Homeopathic medicines.

Homeopathic medicines may be prescribed as part of your treatment plan. These are highly diluted solutions of substances that are chosen for the applicability to your individual problems. They are non-toxic and have no side effects. When even gentle herbs cannot be taken, or the ones needed would interact poorly with other needed drugs, Homeopathic medicines can be taken with confidence. They are safe for children and they love them.

Before you can understand what Homeopathic medicine is all about, you must first understand what *allopathic* medicine is.

53

The greatest number of physicians in the U.S. practice Allopathic medicine; so many, that it is usually referred to simply as "medicine". Rather than being a distinct school with a particular philosophy about healing, it is a collection of diagnostic and treatment methods, some very old and others constantly changing. Illnesses are classified according to their symptoms and causative agents, and are given names. Once the illness is identified, drugs are used to treat the symptoms or to kill disease agents.

Here is where the distinction begins. The prescribing philosophy in what has become "regular" medicine is usually *allopathic*—that is, treating with opposites. For example, if the patient has an over-acid stomach, an anti-acid preparation would be given. Treatment consists of something that causes a condition opposite to the one being treated. High blood pressure is treated with drugs that lower arterial tension; fever is treated with drugs that lower body temperature, etc.

Homeopathic medicine is based on the use of minute amounts of substances that match the patient's overall condition. It is the exact opposite of conventional medicine in that it uses drugs that cause a condition similar to the one being treated. For example, Ipecac is a substance that causes nausea and vomiting. However, if given in very small doses to a person who is sick, it provokes a reaction in the body to relieve the nausea. In the same way, large doses of Peruvian bark extract produces symptoms of malaria; Homeopathic doses of it cure malaria. This is because of natural responses in the body that were observed two hundred years ago by Homeopathic doctors, and only now recognized as the "immune system".

Many substances first introduced by Homeopathic physicians. Nitroglycerin and digitalis for heart conditions, for example, have come to be used by allopathic doctors as well. Homeopathy was brought to the U.S. from Europe to combat several epidemics, such as yellow fever and influenza. Early Homeopathic approaches to scarlet fever, dysentery, and meningitis were dramatically successful.

The cornerstones of the Homeopathic approach are: treatment based on the similarity between the medicine and the person (not based on the disease's name), one medicine to heal the whole person (rather than different drugs for each problem), and the minimum effective dose (often impossibly small amounts of natural substances, which have no toxic side effects).

As you can see, Homeopathy is a philosophy and a method of prescribing medicines, which is why many different types of doctors become homeopaths: MDs, DOs, NDs, OMDs, DCs, etc.

Classical Homeopathy: Three Principles

The first principle or tenet is *similia similibus curentur*, which is a Latin phrase meaning "like should cure like". Each person shows symptoms of his body, mind, and spirit when he is sick. Some of these symptoms are common to the particular sickness and some of them are unique to that person in his sickness. The Homeopathic physician matches the symptom picture of the Homeopathic remedy to the symptom picture of the person, with particular attention paid to those symptoms that are unique to that individual. Thus, for the Homeopathic remedy to be curative, the symptom picture of the remedy

Our pharmacy contains every Homeopathic medicine listed in the standard work, *Boericke's Materia Medica*. Anything that a patient might need is kept in stock.

must be like that picture that the sick person shows.

The second principle of homeopathy is the single remedy. It would be quite impossible for one to know which ingredient was doing what to a sick person if that person were given a medicine that was a combination of ingredients. Therefore, the classical Homeopathic doctor gives only one medicine at a time to the sick person. The doctor allows sufficient time to pass to observe the effects of that one medicine on the ill person.

The third principle of homeopathy is the minimum dose. Drugs given to people in material doses are frequently found to cause side effects or adverse reactions. To minimize this problem, the Homeopathic doctor gives the smallest possible dose so as to maximize the beneficial effects and eliminate the side effects of the medicine. Sometimes the concentration of the substance is less than one in a trillion! Now you see why there are no toxic effects.

Modern Homeopathy

There are non-classical ways to apply Homeopathic medicines as well, and are used extensively in many parts of the world. Rather than laboriously finding the one "perfect" overall medicine for the individual, apply it and wait for the patient to go through a long process of healing (not unlike psychoanalysis), another approach is to use medicines according to the conditions present in the patient. Many doctors in Europe give microdoses of disease agents that caused a particular problem, along with highly diluted "drainage" remedies that encourage the body to eliminate the causative agents. This is a different, but highly effective, way of using Homeopathic principles and medicines.

Another non-classical method is using compound medicines that are formulated for a specific illness, rather than the person who has the illness. Most Homeopathic pharmaceutical companies make these "combination remedies" for various conditions, enabling laypersons to treat themselves. A government study a few years ago found that these were just as effective as the classical single-remedy prescription, which makes many Homeopathic doctors (but not me) mad as a hornet!

Classical homeopaths generally have high intellects, and they enjoy using their brains, as well as huge, old, musty, leather-bound books from a century ago. Think of Sherlock Holmes solving the mystery from tiny clues anyone else would have missed. That is Homeopathic medicine as it has long been practiced. Because of the effort involved, and the long tradition of doing it that way, these doctors are frequently upset if anyone suggests a way to streamline the process or to deviate even slightly from the time-honored methods. I once gave a presentation at a conference about the history of using technology and different instruments to verify the correct prescription, and was never asked to speak there again. In fact, I am not even invited to attend anymore! Yet they argue among themselves constantly about whether to use low potency medicines or high ones in a particular situation, which is the best way to analyze cases, etc.

Whether a doctor uses a low potency or a high potency medicine, a single remedy or a compound, I have found that using electro-dermal screening (EDS) helps tremendously when trying to find a Homeopathic medicine that will help a patient. This simple test lets the person's body tell me what is compatible and effective. Rather than only analyzing the case from a theoretical standpoint, then giving the patient the medicine and waiting to see if it works, I can avoid ineffective measures by testing first. In over thirty years, I have seen thousands of cases where the test has found the effective treatment while the one that looked correct in the textbook proved otherwise.

My Approach

I was trained as a classical homeopath and have a C.C.H. (Certified in Classical Homeopathy) certificate, which is not easy to qualify for. So I am not a "dabbler" where it comes to Homeopathy. Yet, I started to notice over the years that people were not responding to the classical Homeopathic prescriptions like they used to. I used computer programs to help me analyze my cases better and EDS to help confirm choices, but the dramatic cures were still not happening as often as they used to. It became obvious that the current environment is so poisonous and stressful that people with chronic health problems now have layers and layers of toxicity. Getting well is not going to happen without shedding those layers, and they provide obstacles for even the best-chosen Homeopathic medicine to work effectively. I have seen many times how a well-indicated remedy would not get the person better all the way, or might not act at all. After doing some detoxification on the person, I would give the same medicine again, and it would work miraculously! Same person, same medicine. But there were now no obstacles. This follows the concept of healing I was taught as a Naturopathic doctor.

Our clinic has over a thousand Homeopathic medicines in our pharmacy. I use all potencies from the low to the very high. I apply them in a variety of methods, from drainage remedies to the classical prescription, but always according to the case being treated. At one stage of a person's healing they may require one approach; then later, another. By individualizing a person's case I can insure that healing will take place in the most efficient manner. I am not chained to one methodology, and I feel that this is one reason we have such a high success rate at our clinic. In the next section, I will give you a better view of how I use natural medicines in my practice, and why.

The Neo-eclectic Method

Instead of giving only the standardized extract of a botanical medicine, or the highly diluted Homeopathic form of that plant, I have developed over the years a method that is somewhat unique. I call it the *Neo-eclectic* methodology, and high-quality medicines for that style of treatment have been painstakingly formulated. You could think of it as the meeting point between herbal medicine and Homeopathic medicine. Here is the rationale:

The many naturally occurring compounds in a plant can work together (called a *synergistic* effect) to help normalize organ functions. Because of this, whole plants are used in Neo-eclectic medicines; not simply isolated molecules. In addition, the bulk of commercially made botanical medicines contain one herb. Just as one molecule is not sufficient for the health-restoring effects we want to create, one plant may not be enough. We have found that reliable results are gained from taking an exacting combination of herbs in different quantities or proportions, and at different times during the treatment of a case. Neo-eclectic medicines are devised to be as "specific" in their action as Lloyd Brothers' medicines once were, and are prescribed in the same vein as the Eclectic doctors of yesteryear.

Just as a single herb can contain a dozen balanced compounds, a combination of herbs can have many dozens of them; giving more than one combination can increase the likelihood of a synergistic effect that is stronger than the collected conditions that make up the person's illness. A Neo-eclectic compound may be made for each stage of a patient's healing journey, each contrived to cover as many aspects of the case at that time. Like the Eclectic and Homeopathic doctors have historically done, all aspects of a person's case are taken into account, and the dose of the medicine is so small that there are no toxic effects.

A Neo-eclectic medicine is chosen by specific signs, indicating certain tissues are affected in a certain way by the person's disease process. Many plant medicines are known to not only concentrate their action on a specific area, but can also do double or triple duty by being effective for more than one condition. For example, *Carduus marianus* not only predictably acts on the liver itself, but has an action on several different liver diseases. This is because plant medicines tend to restore whole organ systems in the body.

Contrast this with synthetic drugs that are given for single symptoms: those that are anti-inflammatory, anti-depressant, anti-hypertensive, etc.

Neo-eclectic medicines bridge the gap between the need for quick and reliable solutions and the desire for gentle and natural therapy.

DIETARY TREATMENT— The use of foods as "medicines" is a concept common to Eastern approaches to health. According to your health problems, some foods may be important to emphasize and others to avoid. This approach to diet makes use of the fact that a food that is healthy for one person may not be for another.

There is probably no more confusing aspect of health and healing than diet. Every year another book, another seminar and another "one size fits all" approach to diet catches the attention of the public and everyone starts eating high protein, then no protein, high carbohydrate, then low fat, then no fat...and every theory seems to make sense. Until it doesn't work. Or you read something new that seems to make more sense.

Well, we know how lousy those "one size fits all" solutions are in medicine. Why should they be any different in diet and nutrition? And when are we going to realize that those "experts" (you know, the guys who say the opposite thing ten years later) are just getting us to buy into what they're selling right now?

And yet every new approach to diet has its success stories. Why don't they work for everybody? Simply this: Not everybody has the same exact biochemical makeup, so food is going to affect him or her differently.

Think about the Eskimos, who eat mostly whale meat and blubber. Why do these people have so little heart disease? The Scottish have a predominately high fat diet and they have plenty of heart disease. The French, on the other hand, eat a high fat diet also, but counterbalance it with fresh vegetables and lots of wine. The result? Low rates of heart disease.

Calories? The Chinese eat 20% more calories than Americans yet weigh 25% less on average! It isn't just what you take in, it's what you do with it. One's external activities are part of that (harpooning whales tends to burn off a lot of calories), and one's internal activities are a big part of it, too. Internal activities are the processes by which your body uses the fuel you give it. These are determined by several different factors.

59

Ethnic background is one factor. What your ancestors ate for generations is there somewhere in your DNA. But this doesn't mean that you should eat like them, any more than it means that you have the exact same body as all of your relatives. It's a factor, but not a deciding factor.

Blood type is a more individualizing factor. Different people in the same family will have different reactions to foods even though they share the same DNA. Their blood types can predispose them to run efficiently on certain foods and not others. But lots of people share the same blood type and have different bodies and different health problems.

Metabolic problems and organ dysfunction is another important factor. When there are biochemical disturbances internally, your food is not utilized the way it could be. These can be determined by simple testing.

Another factor is food sensitivity. Intestinal allergies are quite common and just because you don't break out in hives does not mean you don't have them. Often, certain foods will break down poorly or cause poor functioning internally. Food sensitivities can be detected by Electro-Dermal Screening (EDS) and corrected by natural means.

Finally, the really big factor is constitution. For centuries, different cultures and the medical methods that sprang up within them have lectured about the different constitutional types. This knowledge has been lost in modern medicine. The different categories of metabolic types, body types, mineral types, miasmatic types, and endocrine-dominance fall into three divisions. Are you

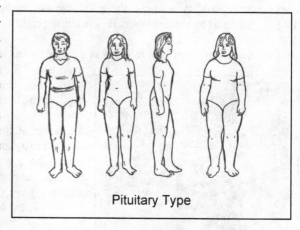

Pituitary Type

a sympathetic nervous system-dominant person (which would make you a slow oxidizer), or a parasympathetic dominant (fast oxidizer)? Or a mixed type?

Are you a cold, hot, or bipolar type? Then, are you hydrogenoid, oxygenoid, or carbo-nitrogenoid?

Are you a positive, negative, or bipolar miasmatic constitution?

Is your morphology (body type) pituitary, adrenal, gonad, or thyroid-dominant?

Is your endocrine system supra-, estro-, para-, or neuro-dominant? Do you even know what I am talking about?

Don't worry. We have made a long study of all these factors and the more of them we evaluate in you, the more accurate is our assessment of you. If we used one or two factors, we would get some useful information and some not-so-useful information—just like a computer search. If we use a lot of key words, we will get what we are looking for a lot faster.

Let's say we have two people with arthritis. They are both overweight. One is always chilly, generally feels worse in the cold, has slow thought processes, has problems with his calcium metabolism, and carries most of his excess weight in his hips and lower abdomen. The other is a woman who is always hot, feels worse at low altitude and in damp weather, and carries her excess weight in her face and upper arms. Do you think these two people need to eat the same things? Do you think their arthritis is going to respond to the same things? Will they lose weight by following the same diet?

The answer to all these questions is "no". On the subject of the arthritis, many natural health authorities advise against certain foods for this and other diseases. A common substance arthritic patients are told to avoid is nightshade plants (tomatoes, potatoes, eggplant, peppers, etc). Apparently, the naturally occurring chemicals in these plants have been known to increase inflammation in arthritic joints. However, they don't *always* do that. Some writers go so far as to state that they "cause" arthritis. Why everyone in the Mediterranean countries, who eats these foods every day, doesn't have arthritis is beyond me! This is another "one size fits all" explanation, and just because it comes from the natural medicine camp, we shouldn't assume that it's true. Only individual testing will show whether a certain person is reacting positively or negatively to a food. By the way, I happen to be one of the unlucky people who don't tolerate nightshades well.

These are just examples of the parameters we use when evaluating someone's dietary needs. In addition, specific foods have been known to have predictable effects on certain organs and processes in the body. Depending on what has been found to be the primary problems with your health, I will suggest foods to use as "medicines" for a time, to have a corrective action on the imbalances that we find. Later, a different diet may be followed for health maintenance. But, like all other treatments, it should be individualized to be the most effective.

FASTING—We discussed detoxification in the early part of this book. The most powerful detoxifying procedure is not something you swallow, like an herbal medicine, nor something that is done to you, like hydrotherapy. It is something you *do*. It is something that is instinctive and is natural for all living beings to do. Do you have a dog or cat? When your pet is sick, what is the first thing it stops doing? That's right. It stops eating. It fasts.

61

Fasting is simply cleaning the body of excess waste and accumulated toxins. This is achieved by resting from food intake and drinking fruit and vegetable juices and/or water.

If you compare your body with your automobile once again, you will see how important periodic maintenance is: if you never replaced worn parts, clogged hoses, etc., never changed the oil or flushed the radiator, your car would break down. When checked and maintained regularly, a car runs better, doesn't break down as often, and lasts longer. The same for your body. This type of upkeep is especially important in today's world, where there are more toxic substances in the air, food, and water than ever before.

Fasting has been used for healing throughout the entire history of mankind. Hippocrates, the Father of Medicine, endorsed it, as do the most scientifically oriented clinics and sanatoriums of today. In Europe, where natural healing methods are as respected as the most "high-tech" medical techniques, there are many facilities devoted to fasting therapy. The Buchinger Sanitarium and the Zabel Clinic in Germany, the Bircher-Benner Institute and the Karolinska Institute in Sweden—these are some of the top names among medical centers which have not only prescribed thousands of fasts, but have conducted scientific research on the benefits of fasting.

Formerly there were facilities like this in the U.S.—the Kellogg Sanitarium of Battle Creek, Michigan being one—but the rise of synthetic drugs and invasive medicine pushed out those clinics that relied on gentle, natural means. In fact, there has been much propaganda that fasting is "dangerous," and the average physician today simply doesn't think to take advantage of this powerful, effective tool.

How Does Fasting Work?

Fasting stimulates healing because it induces a process called "autolysis." Autolysis is a self-digesting metabolic process, one in which the body begins to break down its own tissues and use them for fuel. In other words, since the body is not getting its fuel from without, it begins eating itself! This doesn't sound like a very healthy thing to occur, but just consider this: it is a physiological fact that the body first decomposes and uses for fuel *only those substances/tissues that are toxic waste or diseased tissue*. In the event that there is no more irregular or diseased tissue to consume, the body will only then begin to burn healthy tissue, *but only that which is non-essential for life*. So, self-preservation is an instinct that every cell of the human body is programmed with. People simply do not starve to death from therapeutic fasting.

Now consider this. Disease generally occurs because of a weakened defensive mechanism and/or inadequate functioning of the internal organs. If the rules of maintenance of the body are ignored, metabolic and glandular activity changes, and the body breaks down more easily. In particular, the eliminative organs do not function as they should, and metabolic wastes remain in the body. Toxins from without as well as from within become deposited in the tissues.

62

During a fast, the eliminative organs—liver, kidneys, lungs, and skin—increase their cleansing functions. Tremendous amounts of accumulated wastes are expelled in the urine, perspiration, bowel movements, and even breath of the person fasting. The digestive organs receive a rest during this period, and assimilation of nutrients is generally improved after a fast.

When there is imperfect elimination and abnormal metabolism within the body, waste substances undergo chemical changes. One of these is carbonic acid, which crystallizes and deposits itself in the tissues. Uric acid is another, as well as chlorine, cholesterol, and calcium carbonate. They all form cellular obstructions, and the connective tissue and nerve coverings are the first to receive these deposits. The blood vessels and glands follow suit, as do the bones and organs.

You have probably heard the term "osmosis" used before. It is the process in which a fluid substance mixes with another fluid by passing through the porous membrane separating them. You are aware that blood carries nutrients to the cells and returns with their waste products, carrying them to the eliminative organs. Osmosis is the process by which this happens; the blood does not make direct contact with the cells but communicates through porous membranes. Also, the endocrine glands send their secretions to the bloodstream the same way. So the chemicals that control our metabolism and organ functioning, as well as the nutrients that build our tissues and power our bodies, depend on the passage of fluids through tiny pores. *Deposited waste products block those pores to various degrees*, retarding osmosis and interfering with cell metabolism. Are you beginning to see why fasting is therapeutic for such a wide range of health problems, since so many illnesses could be at least partially caused by defective osmosis?

People who eat a lot of sweets and starches will commonly have a buildup of carbonic acid. It undergoes crystallization after a time and has an affinity for depositing itself in the muscular tissues. Carbonic acid deposits can commonly be found in cases of rheumatic pain.

Heavy meat-eaters typically store up uric acid. Proteins in general, if taken in quantities beyond what your body really needs, will leave destructive residues. Uric acid also crystallizes, and has an affinity for the extremities (especially for the lower extremities, as in gout) and nerve coverings. When it crystallizes, its structure is almost like finely ground glass. I saw it actually come to the surface of the skin of my grandfather, who suffered from gout. During one bad siege, uric acid crystals emerged from this thumb!

Calcium is a cement-like substance. Calcium carbonate, when it forms, has a similar effect within the body. Abnormal distribution of the calcium molecule "cements" tissue; when it occurs in the soft tissues around the joints, it is interpreted as rheumatoid arthritis, and if within the joints and bone surfaces, it is called osteoarthritis. When it occurs within the blood vessels, it is called arteriosclerosis. Faulty assimilation and distribution of calcium causes the substance to be drawn from the bones and deposited elsewhere. Thus, the bones become weaker and other tissues become "cemented."

63

Cholesterol has received a lot of attention in the last thirty years or so. This substance is a very sticky metabolic by-product that has an affinity for blood vessel walls and the heart. When combined with minerals, it becomes arterial "plaque". Medical authorities constantly point out the effects of deposits in the arteries, but almost never mention the tiny capillaries of the body, where it prevents good osmosis. Thus, the exchange of nutrients and waste is hampered by the coating of this substance, and reinforces the whole vicious cycle of imperfect metabolism, buildup of waste material, formation of deposits, etc.

Most people know that they should watch their salt intake and have a vague idea that excess salt can hurt the kidneys and raise the blood pressure. What they don't know is that sodium chloride—salt—is broken down in the body into its two components. Sodium is used to neutralize acids from the food eaten. Chlorine is stored in the connective tissues and in the skin just below the surface. Because it tends to rise in this region, chlorine deposits are easily recognized by thin, shiny skin towards the ends of the extremities. Older people will often have this kind of skin on their hands and feet.

During a fast, carbonic acid has been observed to chemically change to sugar. Uric acid has reverted to protein and cholesterol has been re-converted to fat! Now you can see why fasting is such a potent method for reducing illness and promoting health.

Is Fasting Safe?

In Sweden, two events were held which brought fasting to the attention of the world. In 1954, and again in 1964, groups of test subjects not only fasted for ten days, but also walked three hundred miles during the process! They covered about thirty miles per day, were medically monitored throughout the whole march each time, and satisfied researchers' curiosity regarding the effects of total fasting under stressful conditions. Normally, fasting is conducted in a setting conducive to healing, such as a clinic. Yet, medical tests showed no harmful effects from the fast, in spite of extremely stressful conditions. Blood tests and vital functions were all normal, and continued to show normal in follow-up monitoring months later. This caused physicians to rethink their concepts of starvation, and between the 1954 test and the one in 1964, the ranks of biologically oriented doctors swelled in Europe.

Fasting clinics have recorded cases of patients fasting on water only for ninety days, and on juices and water for two hundred forty-nine days! These figures are the highest, of course, and are not common. But they serve to illustrate what the body is capable of. The average clinic-monitored fast for treatment of a chronic illness is forty days, but most fasts are one-, two-, and three-week fasts.

While fasting with nothing but water should only be performed in cooperation with a physician, juice fasting is a safe, easy process that one can undertake with confidence. For most purposes, juice fasting is just as effective as water fasting. Juice fasting is now the most common method in Europe, for these reasons: fruit and vegetable juices are packed with vitamins, minerals, and enzymes which aid the body in healing and are taken into the blood easily without burdening the digestive system; juices provide the alkaline elements which are valuable during fasting since the tissues and blood are releasing large amounts of acids; raw juices give enough nutrients (without wastes) that the body doesn't become fatigued.

Death from starvation is not possible until a person loses at least one-third of his or her normal weight. If you are contemplating a fast, avoid those people who will tell you that you will die, etc. It is important that your state of mind be positive during this time. Don't tell negative-minded people what you are doing. Those who must know should be told that you are on a "liquid diet." It is amazing how people will respond negatively to an unfamiliar health practice, but will be supportive if it sounds like you are on a weight-loss program!

Contra-Indications For Fasting

If you have cancer, tuberculosis, diabetes, lead poisoning, or a general wasting disease, you should only pursue fasting under the direction of a doctor trained in natural medicine. Such conditions require careful adjustment of the methods commonly used. People who are emaciated from illness should not fast for more than three-day intervals, with a nourishing diet in between; a series of short fasts until the body is built up is the best way.

How To Begin The Fast

It is best to prepare for a fast by going on a raw food diet for 2-3 days first. Meals of uncooked fruits or vegetables (not together in the same meal, however) are a good way to prepare the body for a good cleansing. Apples are a good food to eat in large quantities because they act like brooms in the bowels and clear away debris.

The night before starting the fast, don't eat dinner. Take an enema according to the instructions supplied later in this guide. You will take an enema every evening during the fast. This is a very important part of the cleansing procedure, as it clears large amounts of waste matter from the intestines in an efficient manner. Because manual cleaning of the colon speeds up the elimination of toxins from the body, you feel better during the fast and accomplish more healing in less time.

If you have ever gone a long time without eating and felt bad from it, it was because your body was releasing toxins to be eliminated. As soon as you stop eating, the body steps up the process of pushing out poisons. You want to aid that as much as possible. Now think about the people in those stories you've heard were shipwrecked or lost in the woods and "starved to death" in a few days. Most likely they became weak from the gradual surfacing of toxins and lack of fluids to help carry them off. This, coupled with the fear and hopelessness, could cause them to give up and actually die.

Because *you* are actually directing this process, and making it work as efficiently as possible, you can expect great benefits, but *only if you do it correctly*.

Daily Schedule During A Fast

1. Drink fruit or vegetable juices every few hours as you like, but don't mix fruit and vegetable juices at the same sitting. It's best to alternate.

2. Freshly squeezed juices are best. Avoid commercial brands of bottled juices that contain sugar, coloring, preservatives, etc. Buy your juice from a natural food store.

3. It is best to dilute fruit juices half-and-half with water.

4. Don't forget to drink a lot of good quality water in between juice "meals." The ideal is one glass every 1-2 hours.

5. Avoid especially strenuous physical and mental work, but otherwise follow your usual routine.

6. Herbal teas (non-caffeinated), especially peppermint, are good to drink between juices.

7. At the end of every day, clean the colon with a high enema.

How To Take A High Enema

The average enema bag holds two quarts. Initially, it will be difficult for you to take even a full quart, so begin with a pint of lukewarm (body temperature or 99°F) water. "Bleed" the air out of the tube by releasing the clamp until the water starts coming out of the nozzle. Tighten the clamp and lubricate the nozzle with K-Y jelly, Vaseline, or comparable lubricant.

Hang the enema bag about 2½ feet above you. Kneel on the floor in a head-to-knee position, and insert the nozzle up into the anus. Release the clamp, allowing the water to enter the colon. If you experience any cramping sensations, tighten the clamp, cutting off the flow of water, until they ease. You can control the speed at which the water enters by pressing on the clamp.

Once all the water has been taken, empty your bowels. Then repeat the process, this time with one quart of water. It is good to add some lemon juice to help dissolve mucus in the colon (squeeze ½ lemon, being careful to strain the seeds). Inject the water as before. Now remove the nozzle and lie on your right side. This lets the water drain into your ascending colon. After a minute or so, lie on your back with your knees up. Firmly massage the abdominal region to allow the liquid to reach obstructed areas and to help to release fecal matter that may have accumulated in pockets and convolutions in the bowel. After a minute or two of this, turn onto your left side, letting the fluid drain into the descending colon. After one minute in this position, empty your bowels. Bear in mind that it may take a while for the entire contents to come out. You may have to make a couple follow-up trips to the bathroom later, as everything does not drain at once.

Once a person has become "seasoned" at applying high enemas, a full two quarts can be taken. When gas pockets are contacted by the incoming water/lemon mixture, some cramping results. Just stop the flow of water until the griping subsides, and control the flow by manipulating the clamp, until the rate is a comfortable one.

You will see a variety of matter released from your colon during this process: hardened, encrusted fecal matter, strings or "ropes" of mucus, etc. Even more amazing is the fact that even after a week or more of taking no food, large amounts of excrement are still released! Those with regular bowel movements are astonished to find how much has been retained within the convolutions of the large intestine.

Don't worry about "ruining" your bowels by this process. You will not become dependent on enemas the way some people become addicted to laxatives. First of all, because you are taking no food during the fast, the peristaltic action that begins your bowel movements does not take place. So you must manually empty the bowels to cleanse the body during this period. Secondly, once having resumed eating, the peristaltic action not only returns, but does so more efficiently now that the burden of accumulated waste has been removed. Therapeutic high enemas for a short time pose no threat to your health.

How To Break The Fast

It is important that a fast be broken properly. Short fasts do not require as long a transitional period as longer ones, but still one must take care to ease into eating and not ruin the effects of the cleansing. The major rules are: *Observe a transitional period of 20-30% of the time you fasted* (if you fasted five days, take a day and a half or two days to ease into eating), *do NOT overeat,* and *eat slowly, chewing each mouthful until it is liquefied.*

It is best to begin eating with small portions of fresh fruit. Continue with your juice and water during the first day also. After a couple of meals of fruit, a good-sized vegetable salad should be taken. This acts like a broom to sweep away mucus and accumulations in the stomach and small intestine where the cleansing effect of the enemas cannot reach. The idea is to produce bowel movements before cooked food is introduced again. Take a couple days of raw fruits to break the fast in a healthy manner.

Some other things worthy of note: during a fast, you naturally get hungry. This fades after a couple of days. On a water-only fast, as is sometimes used at fasting clinics, hunger disappears completely after the second day. On a juice fast such as we recommend here, it does not leave as soon or as completely due to the fruit and vegetable juices stimulating the production of hydrochloric acid in the stomach (which creates the hunger pangs). But an interesting thing typically happens: during the fast, one begins craving the worst, most unhealthy foods. Even a food that has been abandoned for health reasons for some time will suddenly come to mind: hamburgers, pastries, etc.

During a fast, you will go through cycles of elimination; waste and poisons will enter the circulation, making you feel sluggish. You will immediately blame it on the lack of food. However, once that load passes the kidneys, you will feel increased energy and well-being. At that point if you check your urine, you will find it loaded with mucus and sediment. Then, the process begins again, removing another layer of toxic accumulations from the tissues and making you feel weak. Persevere, and you will feel even better. It is a known fact that in long fasts, the strength and vitality *increase* with time, rather than decrease! Fasting is an ultimately pleasurable experience, and like any tool, the more you use it, the easier it is to use.

Important: After completing a fast, take a Probiotic supplement to replace the lost healthy bacteria that may be washed away by the colon cleansing. It should contain *lactobacillus acidophilus* and *bifidobacterium bifidus* for good results. These probiotics are available at health food stores or from your health care professional.

What YOU Can Do to Get Healthier

WATER

- **Get a good water filter** and use it for all your drinking and cooking water. This is the first step you should take in getting better. It makes no sense to try to eat healthier and still drink tap water, with its health-destroying chemicals. Also filter your bathing water. **Ask us**—we can recommend some excellent filters. Keep in mind that the filters you commonly see advertised don't take out all the bad stuff.
- **Drink 6 glasses a day**, at room temperature. This is in addition to any juice or teas.
- **Drink a minimal amount of liquid with meals.**
- **Do not substitute distilled water** for drinking water, except for short periods as part of a cleansing regimen.

FOOD

- **Chew your food well.** Liquefy it before you swallow it.
- **Eat consciously.** Don't do something else while you're eating.
- **Stop eating before you are full.** You can have additional fun eating later.
- **Have vegetables at every meal. Eat a GREEN vegetable every day.**
- **Try to have a little fruit every day.**
- **Eat organic foods** whenever possible.
- **Eat a mid-sized breakfast, bigger lunch, and smaller dinner.**
- **Limit refined carbohydrates.** Try not to eat anything white in color. These potent sugars and starches make the body produce excess insulin, which is toxic. This decreases immune function, and leads to increased fat storage.

COOKWARE

- **Never use uncoated aluminum cookware.**
- **Enamel-coated metal, glassware, porcelain, or good stainless steel** should be used.
- **If you use iron pans, don't use for everything.**
- **Teflon-coated pans are passable ONLY if there are no scratches.** If there are, do not use.
- **Do not use aluminum foil in cooking.** It can oxidize microscopic amounts of the metal into your food. Foil is safe to use when storing cooked foods, however.
- **Never eat microwaved food.** Really.

EXERCISE

- **Do some form of exercise every day.** It's not the intensity but the regularity that counts. Walking, Tai Chi, calisthenics, jumping on a trampoline, Yoga, swimming—what appeals to you the most will deliver the best results as you're getting started.
- **It's worth twice as much to you if it's done in fresh air and sunlight.**
- **Stretch every chance you get**, even when you're not exercising.

- **Practice deep breathing**, even if your exercise does not force you to breathe hard. Take ten deep breaths several times a day, expanding the belly first, then the chest.

SLEEP
- **Don't sleep with a light on.** It can interfere with brain hormone production.
- **Review your day and make affirmations that you will make the next day better** just before going to sleep. Since your mind's going to be working while you sleep anyway, you might as well program it positively.
- **Sleep more hours in the winter months**, as Nature dictates.
- **Go to sleep when you are tired, be productive when you are not;** but be honest with yourself about how you feel. Don't kid yourself that you can do a few more things when you're tired. You'll do a poor job, or not enjoy it.
- **Try not to have the head of your bed near an electric outlet** where something is plugged in and drawing current. Also, if there are metal pipes behind the wall there, move your bed.
- **Avoid the use of electric blankets and waterbed heaters.** Use the heater to warm up the bed, and then shut off before getting in.
- **Don't have a TV in the bedroom.** The screen continues to radiate for a while even after being shut off. **The body is much more vulnerable to electromagnetic fields when asleep.**

PERSONAL AND HOUSEHOLD PRODUCTS
There are so many potentially deadly substances in everyday use today and they are hard to avoid completely. But we can limit our exposure.

- **Refuse silver amalgam fillings.** Tell your dentist you insist on composite fillings.
- **Non-toxic cosmetics and toiletries** are readily available now. Try to replace one unhealthy product at a time. The five most common cosmetic ingredients to avoid: **sodium lauryl sulfate**, **diethanolamine (DEA)/triethanolamine (TEA)**, **petrolatum** (petroleum jelly or "Vaseline"™), **parabens** (methyl, propyl, butyl etc.), **propylene glycol**.
- **Companies that have products with safer ingredients: Dr. Hauschka's Skin Care, Kiss My Face®, Tom's of Maine®,** and **Aubrey® Organics**.
- Use essential oils instead of perfumes and colognes, or buy products from the above companies.
- **Do not use commercial deodorants and ESPECIALLY not anti-perspirants.** Get a good natural deodorant with no aluminum or chemicals to absorb through the sensitive lymph nodes in the armpits. **Kiss My Face®** and **Tom's of Maine®** make good ones.
- **Use natural toothpaste with NO SODIUM FLUORIDE.** Or try ½ salt, ½ baking soda as a tooth powder. Don't forget to floss, too. Rinsing the mouth with hydrogen peroxide in water (1:1) after brushing and flossing is a great way to kill bacteria and prevent plaque.
- **Kitchen and laundry cleaners** are available that do not pollute either your body or the environment. Try *CitraSolv*™ or other products for household cleaning, especially if you have small children or pets.

70

- **Don't use synthetic chemical sprays** in the house or on the lawn and garden. If you knew how many people have horrible nerve diseases from exposure to these things, you would never even walk down that aisle in the store.
- **Avoid dry cleaning clothes with the typical perchloroethylene solvent used.** Have clothes cleaned by the "wetcleaning" or carbon dioxide systems (see www.FindCO2.com).
- **Avoid plastic food containers.** Use glassware instead.
- **When painting**, buy paint labeled "No VOC (volatile organic chemicals)".
- If you can do it, **wood or laminate flooring is healthier** than carpeting.
- **Instead of Styrofoam**, wrap in paper.
- **Instead of chemical fertilizers** in the garden, use composted kitchen scraps.
- **Get a negative ionizer** air purifier to eliminate airborne molds, dust, animal dander, hydrocarbons, etc.
- **Keep non-flowering, leafy green plants indoors** for a natural air purification system.
- **Try full-spectrum light bulbs** instead of regular or fluorescent.
- **Use cell phones as little as possible.** Don't carry it on you constantly unless it is in "flight" or "offline" mode. When placing a call, don't put it to your head while it is ringing the other phone (power is higher then). Get close only after the person answers. Even better, use a headset so that the phone is never next to your head. Try not to use a cell phone in a moving vehicle or when the signal is weak (the phone increases power as it tries to reach a new relay tower). And if you have to have one, buy a phone with a low SAR (specific absorption rate) rating, which you can check.

There are good explanations for all these recommendations and I am sure you are wondering about a few of them (especially microwaved food!). But space does not permit going into detail on all these issues in this little book, and some of the scientific papers explaining these subjects are too technical for a work like this. I am giving you the most simplified picture wherever possible, so you can begin to use this knowledge to better your life *today*.

Do I follow these health rules? Yes. Have I broken them? Yes! Probably every one of them at one time or another. But just like the process of getting wealthier, we get *healthier* by gradually accumulating. Try changing just one or two health habits at a time.

PERIODIC DETOXIFICATION

I am in the position of seeing sick people every day, and I can tell you that most of them take much better care of their cars than they do their bodies. Every winter they will carefully staple plastic over the windows of their houses to make them more energy-efficient, but never think about changing other things they do to make *themselves* more energy-efficient. They buy expensive, nutritionally reinforced, fat-balanced food for the cat, and they themselves eat fast food. Then they complain that good quality vitamins or herbal medicines are expensive. So they buy some at Wal-Mart and wonder why they aren't experiencing the miracle that other people did. Ah well, everyone has his priorities. But the person who earnestly wants to get well will make his or her health *top* priority. Whatever pecking order you started out with, the usual categories of money, family, career, friends, community, education, etc., have to be re-shuffled with health *at the top*. Many people have it at the bottom...until they get seriously sick.

There's a great proverb: "Health is a gift, and disease is something you earn."

Since most people are going to continue to eat poorly and break Nature's laws on a more or less regular basis, there is an alternative to getting sick: Simply do a "house cleaning" of the body every so often, and in between, take good nutritional supplements. It's not quite having your cake and eating it too, but it's as close as you're likely to get.

Here's how to "clean house" in one week's time. First of all, you fast during that time. But you do an accelerated detoxification along with it. The lucky thing about it is that you won't feel very hungry because you will be swallowing supplements that will make you feel full. You will also be getting some concentrated nutrition, so your energy won't drag and your body will be building up while it is eliminating waste.

I like to use Springreen products because they have the best track record in producing a quick and powerful cleanse (at various times and locations they have been sold under the name "Sonne's", "Vit-Ra-Tox", and "Veico Products").

For the best health maintenance, do the seven-day cleanse every seven weeks or so (remember that mysterious Rule of Seven!).

Here's the procedure in its simplicity:

1. Stop eating. This is a one-week fast. If you are especially weak, eat raw vegetables and fruits (not at the same time) for the first two and last two days, with liquids only for the middle three days.
2. Every day, put a heaping teaspoon of Springreen #79 intestinal cleanser powder into a large jar with a lid. Now put in 1-2 tablespoons Springreen #77 intestinal detoxificant. Add 10 ounces water, or fruit or vegetable juice of your choice. Cover with the lid and shake it up thoroughly. Drink quickly and follow with a glass of water. Do this 3-5 times a day for the week.

72

3. Optional: Three times a day, swallow five or six Springreen #33 tablets. It is a balanced nutritional food extract made from concentrated juices of cereal grasses. It will keep your nutrients coming in and help you feel full.
4. Every evening, take a high enema (see page 56). Even better, buy a "Colema Board" which enables you to do your own colonic irrigation—much more thorough and easier to do than a simple enema (available from Colema Boards of California, 1-800-745-2446). I have personally used one for over twenty years and they are great.
5. Ease back into eating on the eighth day, as indicated in the previous section on fasting.

If you absolutely cannot fast or do the colon cleansing as described, just do #2 and #3 for a month, drinking the mixture twice daily. This is a good start and will likely begin to correct problems without a rigorous regimen. After a few of these type of cleanses, you will be more confident in trying the seven-day cleanse.

I feel sure that once you experience the amazing benefits of just the simplest procedures like the one above, and some common-sense changes in your lifestyle and personal products as suggested earlier, you will be motivated to do more and more to heal old problems and prevent new ones. We are all boosted by our little successes, and once you see how much better you can feel, I'm confident you won't want to stop there.

When you reach the limits of what you can learn or do on your own, you may want to consult a professional provider of natural medicine. Keep in mind that each such person may have a slightly different approach and may not use all the different methods that I do. The various providers may come from different training backgrounds and have different letters after their names. What is most important is that you find a doctor who has experience and integrity.

One last thing about the "expense" of getting well with natural treatment. I made a quick list of some typical procedures and products in the conventional medical world, and compared them with their counterparts in our world (2008 prices). I think you'll see which approach is affordable. Of course we have to work on getting the price of thermograms down…

Price of Conventional Procedures/Products		Price of NATURAL Procedures/Products	
Adrenal stress index saliva test	$85-100.00	Adrenal (Koenigsberg) urine test	$25.00
Thyroid panel w/ TSH	49.99	Basal metabolic rate test	Free
Thyroid activity assay	24.53	Iodine absorption test	Free
Arthritis panel	42.80	Saliva + urine tests for arthritic case	37.00
Mammogram	87.86	Breast thermogram	110.00
Electrocardiogram (EKG)	100.00	Oscillometric cardiogram	11.00
Spirometry	100.00	Spirometry	11.00
Oral glucose tolerance test	20.00	Total sugars urine test	2.00
Test for *chlamydia pneumoniae*	400.00	Electrodermal screening	60.00
Cholesterol drugs (per month)	80-135.00	Gugulplex (per month)	36.50
21-day course of Amoxicillin	174.00	21-day course of Citricidal Plus	20.00
10-day course of Cefuroxime	123.00	10-day course of Berberine	24.00
Prilosec 100 capsules	400.00	Gastrex 90 capsules	20.00
Celebrex 100 capsules	130.27	Advanced Joint Complex 60 capsules	23.00
Norvasc 100 tablets	188.29	CHF Complex 120 capsules	35.00
Xanax 100 tablets	136.79	Sedaplex 60 capsules	18.00

(2008 prices)

How You Can Treat Acute Illnesses

"Stuff a cold and starve a fever." Remember hearing that old saying? Today, we have a couple generations of people who have been trained to think of illness as an annoying interruption to be treated with a drug so that everything can go on as usual. It is almost a forgotten concept that you should *do* something rather than *take* something.

The mistreatment of fever is one of the most common but serious blunders that people make, one that affects their health negatively. This short explanation will tell you why.

What happens when an infection creates a fever reaction? The invading microorganism has a substance called an *exogenous* (external) *pyrogen* (fever producer). This means that germs have, as part of their makeup, a trigger for the fever reaction. The body reacts to this by releasing an *endogenous* (internal) *pyrogen*. It is a chemical called *interleukin-1*, and it travels through the bloodstream to the hypothalamus gland in the brain and stimulates the production of another chemical, *prostaglandin E2* (PGE2). This causes the body's "thermostat" to be turned up. The body now believes that its temperature is too low and this is the reason there is often shivering. The blood vessels constrict, to prevent heat from them to be cooled along the surface of the skin. The sweating response is shut off. A variety of proteins are produced in the liver to increase immunity. The hypothalamus makes the person sleepy, in order to conserve strength for healing. The PGE2 is also released in the postural muscles, which breaks down muscle tissue and accounts for the characteristic aching in the back and legs common to viral and bacterial infections where there are fever and chills present. The muscle breakdown, while it sounds like a bad thing, actually increases the total amount of amino acids in circulation, which are the raw materials for repair and healing, as well as for antibody production and overall energy. It also depresses the appetite.

The depression of the appetite in "febrile (fever-dominated) illnesses" is an important point. Normally, when someone is sick in this way, there is no desire to eat. In modern times, it is common for those who typically eat denatured foods or foods with high amounts of sugars and chemicals to have a false "hunger". They think they are hungry when their bodies are telling them otherwise. Or they just don't like the imposition of missing a meal, and they eat even though they feel lousy.

This is a mistake because the interleukin-1 effectively shuts down the gastrointestinal tract when the temperature reaches 99.5 degrees Fahrenheit. The now-absent intestinal movements are what make digestion possible but also produce the sensation of hunger. So eating when the hunger signal is shut off (even if you don't know it is) is dangerous because the food will not digest properly. Then it will add to the toxic load of the body. In addition to the substances the body has to nullify from outside, toxins from *inside* are now adding to the burden.

At the same time, the increased temperature is creating an internal environment that is not hospitable to the invading organisms. Not only does the heat inhibit them, but also there is a rapid increase in white blood cell formation, and an increase in their ability to neutralize foreign cells. Amounts of an immune chemical called *interferon* are released, while zinc and iron levels in the blood are decreased, which inhibit the growth of bacteria. Fever will increase antibody production as much as 20 times over the normal levels. Fever is the immune system at work.

Why does the body produce all these changes? It is part of the original "programming" of the many organic systems that is the survival mechanism. Effective defenses against illness require these changes. If you want your illness to be processed quickly and efficiently, you must aid, not hinder, the reactions. Humans have survived thousands of years without antibiotics and other drugs because of this mechanism, part of what natural doctors call the *Vis Medicatrix Naturae*, or healing power of nature.

At an internal temperature of 104°F, the germs that cause polio and gonorrhea begin to die off. At 106°F, syphilis and pneumococcus will die. Malignant cells die between 106°F and 110°F, and some cancer clinics use hyperthermia, or intense heating, to selectively kill off abnormal cells. Note that healthy cells don't die until temperature exceeds 110°F!

If you are my age or older, you might remember seeing fever tanks in hospitals when you were a kid. The official name was "hyperpyrexia cabinet", and it was a cylinder that the patient lay in with his head sticking out, while filaments inside slowly baked his body! While the head was kept cool, the person's temperature, blood pressure, and other signs were monitored. These were "official", approved medical devices, and they saved many peoples' lives from infections that would have killed them. Then, when drug companies developed the new classes of antibiotics, they sent their representatives all over the country to let doctors know about the new miracle drugs. The selling argument went like this: "Doc, you're a busy guy and here you're tying up one piece of expensive equipment on one patient for hours at a time, plus nursing staff. Now all you need to do is spend a few seconds signing your name on a prescription for our drug and you can get rid of many more people's infections than you could before, and make more money doing it." I guess it was a pretty good argument because you never see a fever tank in a regular hospital anymore. But that doesn't mean it didn't work.

Following are some instructions on how you can *use* fever to better do the job that your body intended it to do. You can use these methods for acute illness and also to increase immune system strength to better prevent illness.

Hydrotherapy Procedure
For Fever And Infection

This procedure strengthens immune response, expels toxic waste from the body, and burns out infection. Bear in mind that most infectious organisms require a body temperature of over 102° to be killed.

1. Fill the bathtub with hot water easily tolerated by patient. Put the patient in and take his/her temperature using an oral thermometer. Check the temperature every 10 minutes during this procedure. The goal is to make the temperature go **up** and thus assist the body in what it is trying to accomplish with the fever.

2. Increase the temperature of the bath water by adding more hot water, as hot as the patient can stand. Put towels moistened with the bath water on the shoulders and knees if they are exposed.

3. Once the patient begins sweating, put cold compresses (cold wet towels, ice bag, etc.) on the head/face. Give room temperature water to drink, as much as desired. Don't worry! If you keep the head cool, the fever won't do any damage.

4. The patient should stay in the tub for at least 20 minutes (if under ten years of age, one minute for each year of age). After, the patient should bundle up with several layers of clothing and several blankets. During sleep/rest, the patient should sweat profusely. You may change the bedclothes, but don't allow patient to become chilled during changing. The fever will "break" with the perspiration. The patient can drink hot ginger tea to help increase perspiration and break the fever.

Do not perform the above procedure more than 5 times in a week. Do not perform during the first trimester of pregnancy, if you have low blood pressure, heart disease, thyrotoxicosis, tuberculosis, anemia, or blood loss. Be careful with hot water if you have diabetic neuropathy or have decreased sensation in the legs.

Cold and Flu Protocol

One reason a common cold lasts so long is that you are reinfecting yourself through the respiratory tract. The germs in your throat are reproducing at great speed.

1. You should repeatedly drink half-pints of hot fluid (as hot as possible). This will continually kill the viruses (and the bacteria that accompany them) responsible for the continued presence of your cold symptoms. DON'T do it as you usually do, where you take five or six sips of hot liquid, and then drink the rest of it warm. It means 200 or more hot sips or mouthfuls of

piping hot liquid within those first 24 hours. The hot water can be flavored with tea, soup, etc. But if the water, tea or broth is warm or lukewarm, it won't do its job of killing the bugs. It must be hot, must be repeated, again and again and again.

2. In between cups or bowls of hot liquid, DRINK COOL FRESH WATER. Pints of clean fresh water help accomplish primary task of flushing out the body. All the waste products, including killed organisms, have to have somewhere to go!

3. Try not to blow your nose. Let it drain, wipe it, or "sniff and swallow"—but don't blow it out. Forceful blowing will drive infected material back into your Eustachian tubes and sinuses, leading to an entrenched infection. Your body can eliminate the mucus that the bugs are occupying, but you don't want it being driven into the nooks and crannies of your head.

4. DON'T EAT ANY SOLID FOOD for 12 to 18 hours—especially if you have a fever; even a slight one. Digestion uses a lot of energy and a great deal of the body's resources could be used to get well quickly. Allowing the body to focus on getting rid of the infection makes sense; diluting its power by making it concentrate on other functions slows the whole process down.

5. REST—ESPECIALLY YOUR EYES. Your eyes are using most of your energy now. Keep your reading to an absolute minimum for the 12 to 18 hours it takes to eliminate the majority of your cold or flu symptoms. There are numerous sophisticated lenses in each eyeball, and the energy required to use them is enormous. Give them a rest. That saved energy accelerates the healing process!

6. Take zinc, either in liquid or lozenge form. It tastes lousy, but every major study done with zinc between 1979 and 2000 has shown zinc to reduce length of a cold by fifty percent. This also applies to flu as well.

7. If you combine a natural medicine, hot fluids, abstinence from food, rest, and zinc, your cold or flu will be eliminated in record time.

Remedy For A Tickling Cough

A sleep-preventing cough is many times unresponsive to lozenges or cough syrups. Stronger drugs to paralyze the nerves feeding the cough reflex, containing codeine or other narcotics, are harmful in the long run. Apple cider vinegar, on the other hand, is harmless and effective.

Mix two teaspoons of apple cider vinegar in a glass of water and keep by the bedside. Take a few swallows whenever the tickling is felt.
The tickling cough will usually disappear quickly. Repeat as needed.

Note: Do **NOT** use regular white vinegar or any other kind of vinegar for these purposes. Only vinegar made from apples has the characteristics needed.

Remedy For Laryngitis

Acute laryngitis often responds well to apple cider vinegar. One teaspoon of vinegar in a half glass of water, drunk every hour for seven hours, has restored the voice completely in most cases. Meals should be very light, if eating at all (fasting would be better).

Note: Again, do **NOT** use regular white vinegar or any other kind of vinegar for these purposes.

Infinity Health Care, Ltd. main location:
1211 Pineview Drive, Morgantown, West Virginia 26554.
Telephone (304) 296-6606.
Branch offices and clinics may be located near you.

Dr. Negri's web site can be found at: **www.drnegri.com**